Stepping in Wholes: Introduction to Complex Systems

Stepping in Wholes: Introduction to Complex Systems

Topics in Process Adaptive Systems

by Jim Ollhoff and Michael Walcheski

Sparrow Media Group, Inc.

Eden Prairie, MN

Printed and bound in the United States.
First edition, 2002.

Publisher's Cataloging-in-Publication Data

Ollhoff, Jim.
    Stepping in wholes : introduction to complex systems / by Jim Ollhoff and Michael Walcheski. - - 1st ed. - - Eden Prairie, Minn. : Sparrow Media Group, Inc., 2002.
        127 p. : ill. ; cm. (Topics in process adaptive systems)
        Includes bibliographic references (p. 118-123) and index.
        1. System theory. 2. Organizational behavior. I. Walcheski, Michael. II. Title. III. Series.

LCCN: 2002105179
ISBN: 0-9719304-0-6

"This new learning amazes me, Sir Bedevere. Explain again how sheep's bladders may be employed to prevent earthquakes."

King Arthur
Monty Python and the Holy Grail

Table of Contents

List of Charts, Illustrations, and Free Tips

Introduction

C omplex systems is about a thought process; it's a way of seeing the world. This is not a set of techniques; it's not a cookbook approach to organizational theory. Complexity theory is not about a list of tricks and tips to help you create a better organization. Complexity theory is about a different way of *understanding* the organization.

Complex systems is a way of thinking that crosses over all disciplines; and in this book, we're interested in applying systems thinking to organizational life. It's a way of understanding all the strange patterns that go on in organizations. It's a way of understanding so that we can do things so that things actually get done.

Our model, which we have called *Process Adaptive Systems,* is a way to understand the functioning of systems. We have developed this model for people who need a better understanding of organizational life. Our work is based heavily on the pioneering work of the late Edwin Friedman.

This book is an overview of Process Adaptive Systems. The idea here is to get a rough familiarity with the concepts and the landscape.

Foundations of the Process Adaptive Systems model include a consideration of system process rather than content, an attention to adaptive patterns rather than static mechanisms, and a recognition that organizations are more akin to living ecological systems rather than machines.

Process Adaptive Systems draws principles from the disciplines of biology, family therapy, and physics. From the field of biology we draw principles of adaptation as key process in systemic change and the interdependent web of relationships present in all ecology. From the field of family therapy we draw principles of differentiation as a critical process and emotional fields as a motivator for action. From the field of physics we use the notion of interaction as a mechanism for influence and the idea of chaos as the natural state of nature.

Of course, we also make use of traditional systems theory when we discuss feedback loops, organizational learning, and the recognition that analysis (breaking things down into their component parts in order to understand them) is frequently a misleading way to understand the whole.

Complexity theory is really a body of theories. Process Adaptive Systems is one model of integrating systemic phenomenon into a unified whole.

Our objective is to give concepts and principles to people so that they can be more effective leaders and thinkers. We want people to be able to use systems principles for more effective observation, diagnosis, problem-solving, and leadership of organizations.

We hope you enjoy this short introduction to Process Adaptive Systems.

Jim Ollhoff
Michael Walcheski

Part One: A Green Lawn or a Healthy Lawn? A Primer on Systems

Chapter 1

Grass

Most people have lawns in their backyard. For many people, there isn't anything more satisfying than a lush green grassy lawn stretching out across their property. Many people try their hardest to create the lush green lawn. Some succeed, but others look out across their yard and see weeds, brown spots, animal holes, fungus, and dirt.

Grass doesn't just grow by accident. Some people fling some grass seed into their yard and expect it to grow. Sometimes it will. But for the optimum grass experience, the gardener must take into account a variety of variables and be attentive to how they interact.

The authors like nice grassy lawns.

The composition of the soil is critical. Soil needs to hold just enough water so the grass roots can drink, but not so much that the grassy roots will drown in standing water. If the soil is too sandy, it will not hold water very well. If it has too much clay, it will not drain properly. If your soil has too much clay, don't add sand, or your lawn will turn into cement. Whether your soil has too much clay, or too much sand, the addition of organic material will nurture a nice, rich soil called *loam*.

The pH balance of your soil determines whether or not your plants can access the nutrients. If you soil is too acidic or too alkaline, the grass may not be able to use the fertilizer.

The compaction of your soil is another factor determining the health of your lawn. If your soil is too compact, the grass roots cannot penetrate. Roots, like other living things, need oxygen, and if the soil is rock hard, the grass cannot grow deep. If the soil is hard, the roots will push toward the surface, creating shallow root growth,

and thus unhealthy grass. If your soil is too compact, you can aerate—a process that pulls cigar-shaped plugs out of the ground,
allowing the grass some room to breathe.

Microbes are necessary for healthy lawns. Microbes will break up
the grass clippings, the dead roots, etc.,
into *compost*—the organic matter that
feeds the lawn. Worms burrow through
the ground, creating more organic
material (the presence of worms is a
good sign of a healthy lawn). Your lawn
needs a lot of organic material in the
soil. Without organic material, the soil
will turn to dusty cement.

This grass thing is a long metaphor that sets the stage for complex systems and systems thinking.

The right amount of water, of course, also helps to determine the
health of a lawn. Too little water and the grass dries up. Too much
water and the grass drowns. Frequent shallow watering makes for
shallow roots, as the roots will constantly push to the surface to get
the water.

Fertilizing is another ingredient of a healthy lawn. Grass needs a
variety of nutrients, many of which are provided by fertilizer. Be
careful though, with using chemical fertilizers. Besides giving your
lawn more nitrogen than it needs (which will eventually run off into
the rivers, creating an algae bloom, thus using up all the oxygen that
the fish need), chemical fertilizers tend to chase away worms and
microbes. Worms and microbes, like most humans, don't like being
bathed in chemicals, and they will go live somewhere else. Without
microbes and worms, you'll have no way to break down the organic
material, and the soil will become harder and less fertile. We recom-
mend the organic fertilizers. Organic fertilizers deliver long-lasting,
low-dose, non-toxic nutrients.

The presence of thatch (a layer of decomposing organic material
above the ground) is important for thriving lawns. Without that layer
of thatch, water will drain too quickly and parasites can invade your
lawn much easier. However, when thatch gets thicker than a half-inch
it will keep out water, sunlight, and fertilizer, and create a welcome
environment for bugs and fungus.

The right grass seed is necessary for healthy lawns. Different
kinds of grass are more tolerant of shade, drought, heat, winter, etc.,
and so determining the kind of grass for where you live is critical.

Finally, cutting the grass is another factor in determining lawn health. Cut your grass too short, it will be tough for the little guys to survive. If you cut your grass too long, it will be a haven for parasites and fungus (and it will look silly, too). If you cut your grass too infrequently, it will shock the grass when too much is lopped off. If your mower is too dull, it will tear the plants instead of cut them, leaving them open for disease.

A lawn of 10,000 square feet will produce 2,000 pounds of grass clippings each year.

Your lawn is a complex system, with many different parts. If everything in your lawn is perfect, except for the pH, your lawn will not be able to absorb the fertilizer. Then, your grass will wither and die. Bummer.

If one part of the system goes bad, the whole lawn will suffer.

Further, you cannot affect just one part of the lawn system. For example, you cannot stop watering the lawn and hope everything will be okay. Even if you were careful about every other variable—adjusting the pH, removing excess thatch, and providing nutrients—but your grass didn't get any water, your grass would die.

The different grass variables also interact with each other. If your pH is off, you may need more fertilizer (that's "the grass" will need more fertilizer, not "you"). If you use chemical fertilizers, you may need to aerate more often. If you use Kentucky Bluegrass seed, you'll need to cut lower than if you use Tall Fescue grass. If you have a clay or sandy soil, you'll need to add more organic material to the soil.

Kentucky Bluegrass is widely regarded as the most attractive and most desirable grass.

Your lawn is a system. For the best possible lawn, you can't think about just one of the variables. You'll need to consider all of the parts of good lawns. You'll need to consider all of the parts and how they interact with each other. You'll need to be a systems thinker.

Free Composting Tips

Nothing is as good for your soil as good old compost. Compost is decayed organic matter; it conditions the soil and provides natural nutrients. More than half of all the junk that goes to landfills is organic matter. Rather than filling up our already overstuffed landfills, we could take out the organic material and recycle it into the soil. Further, our topsoil is so eroded that we need organic material back in our soil. Few things are as natural, helpful, and environmentally healthy as composting.

And you, too, can do this in the safety and convenience of your own back yard.

To have a good compost pile, you need roughly equal amounts of nitrogenous and carbonaceous material. Nitrogenous material is grass clippings, food scraps, and the soft organic stuff. Carbonaceous material is sawdust, tree bark, leaves, and the more "woody" stuff. Then you need enough water to make the mixture damp, not wet— like a wrung out sponge. With just the right mix, and the occasional stirring, in a couple of months you'll have a stew that will turn to black, soil-like compost. It will have a pleasant earthy smell, and you can dig it into your lawn or garden.

If you have too much carbonaceous material, the compost pile will not heat up, it won't decompose, and you'll just have a pile of woody crap in your back yard.

If you have too much nitrogenous material, you'll get too much decomposition. The wrong kind of microorganisms will take over and generate a terrible, pervasive ammonia smell. Dogs will come from miles away to roll in it, green gasses will billow up from it, paint will peel off the nearby houses, birds will die, small animals will spontaneously combust, and the police will come. Found that out the hard way.

So, if your compost pile is cold and uneventful, you'll want to add more nitrogenous material. If your pile is getting too hot, you can slow it down by adding carbonaceous material. Your compost pile will be whatever you design it to be.

Similarly, when you design a system, you will get the results that you have designed. The system will simply behave as it has been designed. If your system is misbehaving, sometimes, you simply have to structure it differently.

Chapter 2

What is a System?

A system is a group of parts that function as a whole. When you affect one part, you automatically affect all parts. When you affect the whole, all parts are affected. The individual parts are in some kind of communication and feedback with each other.

- A family is a system. What happens to one member of the family will have an impact on the rest of the family. If the father is laid off from work, the rest of the family will need to cut back on spending. When the mother goes away on an extended business trip, the rest of the family has to take over the tasks that the mother normally did. When the child experiences behavior problems at school, the parents will usually spend enormous amounts of time thinking about it.

> *A system is a group of parts that function (and sometimes dysfunction) as a whole.*

- A forest is a system. All the parts interact and contribute to "the forest." If you upset one part of the forest, the whole forest is disturbed. If you take out one link in the food chain, the whole food chain can fall apart.
- Recently, a small African country was having trouble with herds of wildebeests. There were just too many wildebeests for the people settling some of the jungle areas. So, they brought in a pride of lions (lions like to eat wildebeests). However, the lions decided they preferred the taste of wild pigs. So, they had a rapidly expanding lion population, a rapidly expanding wildebeest population, and a rapidly declining wild pig population. The balance of nature had been thrown askew.
- Recently, there was concern in some of the tropical areas of Central America about the possibility of the extinction of a particular species of wasp. The first reaction of many people was, "who cares about that species of wasp?" However, in most

tropical areas of the world, there is only one species of wasp that pollinates one kind of fruit tree. Without that wasp, that particular kind of fruit tree could die out. Without that kind of fruit, the forest animals may do without some kind of nutrient, making them more susceptible to illness. So, the disappearance of one species of wasp could have implications that ricochet throughout the entire forest.

- The human body is a system. If you hit your thumb with a hammer, it is not just your thumb that reacts. Your whole body reacts; you may jump, shout, utter various and sundry words, and flip your hand back and forth. The thumb was the only part of the system that was damaged, but the entire system reacts.

- An organization is a system. There are many different people in an organization, each with different personalities and agendas. The organization has many different units, each needing to relate to each other for the good of the organization.

  We need to distinguish, however, between a collection of things and a system. A collection of things is not a system. In order for a system to be a system, it has to interact for a goal (in organizations, sometimes the true goal is the opposite of the expressed goal).

- Silverware sitting in a drawer is a collection of things, not a system.

- A pile of dirty clothes is not a system (unless it becomes malodorous or spontaneously combusts).

- A pair of tusks, some pachyderm DNA, some skin, hair, heart, lungs, and a trunk, is not an elephant (it's a disgusting mess).

Chapter 3

What is Complex Systems Thinking?

T hinking in complex systems is a way to see the world, looking at wholes and their interactive pieces instead of the constitu-ent parts. Systems thinking is to look at the dynamics of the entire whole (the whole whole?). Systems thinking looks at the whole first, and is less concerned with analyzing the parts that make up the entity.

Systems thinking is not discipline-specific. In other words, the basic principles of systems thinking are applicable to a wide variety of fields and pursuits.

We usually consider the "start" of systems thinking to be in biology in the early twentieth century. In 1928, biolo-gist W.E. Ritter said, "Wholes are so related to their parts that not only does the existence of the whole depend on the orderly cooperation and interdepen-dence of its parts, but the whole exer-cises a measure of determinative control over its parts" (as quoted in Mayr, 1997). Ritter and others realized that everything is related and interdependent. Affecting one part of the environment will affect all parts of the environment. The ecosystem hangs in a balance, and if we disrupt one part of the balance, we can throw the whole system out of whack. This is the principle of *interde-pendence*, and it is applicable not only in biology, but to other fields as well.

Systemic *is a word that means "relating to systems."* **Systematic** *is a word that means, "carried out in an organized manner." The words are different. Don't be fooled.*

In the 1940s, American scientists tried to automate the firing of anti-aircraft guns, so the gun would "take into account" that the airplane was moving. The scientists realized that they had created a machine that "learned" through feedback. This was the origin of *Cybernetics*, a forerunner of systems theory.

Biologist Ludwig von Bertalanffy (1969) was one of the first people to generate a coherent theoretical model of systems theory, calling it *General Systems Theory*.

Family therapists, early in the 1970s, began to think systemically. Virginia Satir (1972) and others realized that a family's "problem" is not always what it appeared to be. Frequently, a family problem erupted in the weakest link. The "weak link" is not the problem; it is simply where the problem manifests itself.

In the 1970s, scientists looked at phenomenon that used to be considered completely random (i.e., weather, soap bubbles, dripping faucets, and epidemics), with complex computer programs, and realized that there was an underlying order. They called this new science *chaos*, and it concerns itself with the patterns behind complex and seemingly "unpredictable" phenomenon.

Complexity science *is sometimes used as an umbrella term for a group of theories that include systems theory, chaos theory, and cybernetics.*

So, we take principles out of biology, family therapy, and physics and apply them to other fields. We call complex systems *transdisciplinary*—it takes principles from a variety of fields and uses them in a coherent way to understand other fields.

With the publishing of *The Fifth Discipline* (Senge, 1990), systems thinking came to organizational management. Most of the thinking about systems in organizations has relied on the concept of the learning organization. Organizations that learn can benefit from their experience rather than being tossed around in the randomness of a rapidly changing environment.

Process Adaptive Systems is one particular "style" of systems thinking. It is a set of integrated propositions that give broad insights into organizational diagnosis, problem-solving, and leadership.

Chapter 4

Why Do We Need Systems Thinking?

S ystems thinking gives us a better picture of what's going on in an organization. When we use the tools and the lenses of systems thinking, we get a more accurate perspective on organizational observation, diagnosis, and problem-solving.

Have you ever experienced any of these in your organization?

- A reoccurring problem that won't go away, no matter what you do to solve the problem.
- Lots of complaining and sour attitudes; from month to month the complaining continues, just shifting the topic of the complaint; there is even complaining about the complaining.
- Organizational secrets—things that you think go on but no one talks about.
- Policies and rules that actually hinder the stated goals and purposes of the organization.
- Organizations that never seem to get anywhere; they just stay busy fighting the daily fires.
- Logical people who sit in a committee and make illogical decisions.
- Organizations that never seem to get anything done.
- A few people, usually the fussiest and most-disliked, seem to control the agenda of the organization.
- The harder the leader works, the less that seems to get done.
- The more an organization tries to change, the more they seem to stay the same.

Process Adaptive Systems builds heavily on the work of the late Edwin Friedman. Friedman built heavily on the work of family therapist Murray Bowen.

- A lot of people expressing a desire for their company to become a learning organization, but no one knows how to accomplish it.
- Whenever the organization tries to change, a crisis happens that stops the change because people have to deal with the crisis.

Probably, you've experienced many of these problems. These problems are extremely common patterns of behavior that are found in most organizations. They are problems that stump most people. In fact, when people try to understand and fix these problems in a traditional mindset, they usually make the problem worse! These problems, however, can be more easily understood—and corrected—from a systems perspective.

Chapter 5

A Different Way of Thinking

S ystems are a different way of thinking than most people are accustomed. Most people think in a straightforward, cause-effect, short-term fashion. This is *linear thinking*. We get bombarded constantly with linear thinking.

- The Five O'clock News can reduce any complicated social problems into a few sentences. In real life, the issues are usually not quite that superficial.
- We constantly feel the pressure to get "immediate results." Unfortunately, immediate solutions generally cause long-term problems.
- Organizations are constantly enjoined to "meet the budget for this fiscal year." They perform extraordinary fiscal gymnastics to be in the black by the end of the year. Frequently, however, organizations act as if next year does not exist—in other words, they act at the expense of long-term sustainability.

Complex systems thinking is frequently counter-intuitive. That means that when you say something systemic, people will think you've lost your mind.

- Politicians lambaste the alternative viewpoint with superlatives like, "it's too expensive," or "it won't do anything," or "it's not what people want."
- Organizational leaders try to motivate their employees by suggesting, "they just need a good kick in the pants." (Will this actually accomplish anything, aside from an assault charge?)

People usually start on the road to systems thinking when they come to impasses in their understanding. They try to understand organizational dynamics, but they never seem to get it right. All the linear thinking tools fail to help them understand a situation or do something about it. When people reach this impasse, many simply quit trying to understand the organization. Others stay persistent, trying to understand the organizational processes. With persistence and unending curiosity, people eventually become systemic thinkers.

Stephen Haines (1998), paraphrasing Russel Ackoff (1991), identifies constructs of thinking in the non-systems perspective. Since the time of the middle ages, we have lived with the concepts of *reductionism* and *mechanization*.

Reductionism is the idea that if you take something apart, and examine its constituent parts, it will reveal how the whole works.

- Physicists used to believe that they could learn how atoms work by looking for the atom's smallest particle.
- Biologists used to believe that they could manipulate only one variable in the environment, with no other effect (like putting down fertilizer to make the crops grow; this action has many effects, some desired, some not desired).

A popular book that details the systems perspective is by physicist Fritjof Capra, called, **The Web of Life.**

- Education consultants used to believe that if the teacher had just the right kind of training (inputs), then their teaching (output) would be just fine. In fact, one of the government committees in the 1970s recommended that elementary school teaching (output) would be improved if every teacher had a PhD (input).
- Organizational leaders believed that if they could just do or say the right things, then the organizational problems would evaporate, employees would be motivated, and positive change would blossom.

When we have a problem, says reductionism, we just "cut the problem down to size" by looking at the different parts of the problem. We analyze the behavior of the parts and then we think that tells us something about the behavior of the whole. This kind of technique is so common, and so ingrained in us, that we think this kind of analysis is equivalent to good thinking!

Mechanization is the idea that every phenomenon can be explained by one simple rule: cause and effect. Mechanization says,

"Every effect stems from one cause." Further, every problem has a single solution. While this kind of thinking sometimes works when you deal with machines (if a car runs out of gas, there is a single problem. It can be solved with a single solution: putting gas in), this kind of thinking does not work very well when dealing with social systems and human interactions. This limits our ability to understand what really goes on in an organization.

- A mechanistic answer to an organizational problem is to blame someone (usually the leader). The mechanistic thinkers say, "If we could just get rid of that incompetent manager, our problems would go away!"
- A mechanistic answer to the high rate of divorce is to make it more difficult to get a divorce. For the mechanist, one effect (the high rate of divorce) has one cause (divorce is too easy), and therefore, it has one solution (make divorce more difficult).
- A mechanistic understanding of youth violence is to say that it's the fault of Hollywood and violent movies. One effect (youth violence) has one cause (violence in movies). While television violence has shown to be *a factor* in youth violence, it is a much more complex problem. To understand youth violence today, we need to understand a variety of influences on the lives of youth: the changes in the socialization process over the last forty years; the changing influences of schools; the effects of parental discussions with children on violence; the child's perception of economic injustice; the child's environment of violence in the home and community; the presence of significant non-parental adults in the child's life; the ability to think through social situations; time spent with parents and other adults; the presence of appropriate neuro-transmitters in the brain; the effects of peers; etc.

Chapter 6

Some Key Concepts in Complex Systems

Interdependence

When a system is interdependent, all parts of the system are affected by all other parts. Systems are, by definition, interdependent. It is not necessarily that "one action has one effect" (linear thinking). It is not even that "many causes have one effect" (this is known as *multiple causation*. Multiple causation is a more sophisticated form of linear thinking). In systems thinking, all parts affect all parts. An action in one part of the system can affect a part of the system that seems to have no connection to the original part. You cannot take an action without having some effect—and the action you take will frequently have an effect that you didn't expect.

Open and Closed Systems

When a system is open, the boundaries are permeable. Information and stimuli flow into the system, and information and stimuli flow out of the system.
- For example, an open system is a lake. Water flows from a river into the lake. Rain falls into the lake. Animals and birds move in and out of the lake. So, the lake takes in stimuli and actions from outside itself. Further, a river flows out of the lake. Lake water evaporates. Fish that are born in the lake swim downstream. So, the lake gives stimuli and actions to the surrounding system. You know that the lake is a lake—you know where the boundaries are, where the lake starts and stops. But, the boundaries do not keep stimuli and action out—there is a free exchange of phenomenon back and forth. This is an open system.
- Your body is an open system. You take in food, air, and water from the surrounding environment. You are influenced and

affected by the experiences, the relationships, and the environments around you. You are influenced by the emotional systems around you. And conversely, you affect the system.

A closed system shares nothing with its environment. It is self-contained. The boundaries of the closed system are extremely rigid— nothing goes in and nothing goes out. Usually, closed systems cannot survive indefinitely. Their resources get used up.

> *Boundaries are the edge of systems. They mark where one system stops and another begins.*

- The textbook example of a closed system is a wristwatch. When the battery runs down, it stops. The watch is sharing nothing with the outside environment, and the outside environment is sharing nothing with the watch.
- An example of a closed system, at least on the metaphoric level, was the old Soviet Union. They kept influences out. They jammed incoming radio signals. They built walls so that the boundaries would be rigid. No one could get in and no one could get out. However, the country deteriorated. Economically, socially, and emotionally, the Soviet Union ran down. Without stimuli and actions entering and exiting, the system cannot survive.
- Some family systems build emotional walls around the family to "protect" the members from outside influences. Sometimes, families that are in grief or in crisis become temporarily closed. This closure helps the family to emotionally cope. Families that are *permanently* closed run the risk of becoming self-contained, deteriorative, and finally implosive.

Closed systems are self-contained. They have pre-determined actions. They run until they run down.

All systems are on a spectrum line between completely open on one side and completely closed on the other side. In real life, there are probably no examples of systems that are perfectly closed or perfectly open. Systems, however, tend to be on one side or the other.

Homeostasis

Homeostasis, sometimes called a *balancing process* or *compensating feedback*, is the work a system does to keep things the same. Systems regulate themselves to maintain a steady state. Deviations from the standard initiate actions to correct the deviation. Homeostasis is occurring when the system is in balance. A homeostatic influence is a mechanism that brings the system back in balance.

In most organizations and other systems, there is a push to stay the same. This is normal and you should expect resistance to change.

• When we are sick, our body temperature goes up. This fever is a deviation that pulls the body out of balance. We cannot survive when our body temperature varies too much. So, our body has some homeostatic tools that it puts into place: perspiration and heavy breathing. The fever was the deviation. The perspiration and the heavy breathing push the body temperature back to normal—back to a homeostasis. Without the body's homeostatic influence, a simple fever would be fatal.

• Kaufman (1980) uses the example of wolf and deer populations. As the deer population increases, the wolf population increases, because there is more food. Then there are too many wolves, so too many deer are eaten. So the deer population goes down. Because there is less food, many of the wolves starve. So, because there are fewer wolves, the deer population goes up. Then, because of all of the food, the wolf population increases. Consequently, over time a balance—a homeostasis—is maintained between the wolf and deer populations.

• Organizations maintain a homeostasis as well. Some organizations require a lengthy approval process for new initiatives. The proposal may go through five committees, each approving or disapproving it. While the committee structure is lengthy and bureaucratic, it prevents risky behavior. Most committees, in fact, serve the function of providing a homeostatic influence.

• When a well-meaning leader comes into the system, and sees that the organization "needs to change," the leader often responds by working hard. However, when the leader overworks, the system tends to "underwork." A balance of work is maintained in the system. The leader, frustrated by the lack of progress, will work harder... and then the followers will underwork more. Homeostasis is maintained. Sometimes, leaders need to strategically underwork to move the system. Sometimes, to get more done, we need to do less.

Figure 1
Homeostasis in Populations of the Wolf-Deer Relationship
(Kauffman, 1980)

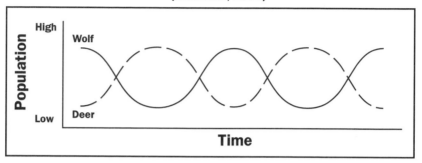

Anxiety

There is a difference between stress and anxiety. Stress is a physiological and emotional arousal based on a threat (real or imagined).

- Stress is the panicked feeling we get when someone cuts us off on the freeway.
- Stress is the burdened feeling when we stare like a deer in the headlights at a "to do" list that is way too long.
- Stress is the bad feeling we get when the boss yells at us.

Stress usually comes up quickly, and passes in a short time. When we have stress, our heart beats faster, muscles tighten up, adrenaline pours into our system, and the blood leaves our skin to move to the important organs (leaving us cold and clammy). When we are under stress, our body moves into a mode where we are ready to fight or run (this is called the *fight or flight syndrome*).

Anxiety, on the other hand, is long-term and enduring. It is often a learned response from conflict in our family of origin. Anxiety may stem from current conflicted interpersonal relationships. It may stem from a fundamental perception of a lack of self-worth. It may stem from a sense of impending doom in the organization. Stress comes and goes, but anxiety is long-term.

Stress is a fear of "what is." Stress occurs in scary situations— we're caught in a traffic jam or a bad performance review. Anxiety is a fear of "what could be." Anxiety even occurs in non-scary situations because we're afraid of what might happen.

Anxiety builds up in people, and they carry it with them into the systems with which they interact. It can't be seen, and you can't put your finger on it and say, "there is the anxiety." People find many ways to hide anxiety or cope with anxiety. You can't "see" it in people, but you can see it in the space between people—in their interactions.

- The boss approaches an employee because the employee made a poor decision. In the employee's family of origin, the father would scream and shout at the poor decisions of the family members. The same feelings of anxiety that the child felt, reoccur in the employee when the boss discusses the poor decision. This is anxiety.
- In a particular family, children learn never to mention conflict. In that family, it is better to pretend that everything is okay, rather than acknowledge the conflict that everyone feels. So, a child from that family grows up and gets into a management position, where he covers up and glosses over any type of conflict. When conflict does occur, the manager has anxiety.
- A stepfather physically abuses a child. The child grows up, and when people approach the adult in the same way the stepfather did before a beating, the adult has a panic attack. This is anxiety.

In the previous examples of anxiety, all of the people will have the same physiological reaction—increased heart rate, muscle tension, cold and clammy skin, etc. Anxiety will have the same physiological reaction suffered by people undergoing stress. Anxiety, however, will have an additional emotional component—feelings of worthlessness, shame, fear, or heaviness. Anxiety strains our ability to cope, so we need to find ways to dilute it.

Sometimes anxiety in a system becomes so overpowering that it begins to influence people's actions. It can even become noticeable to outsiders. We call this an emotional climate.

Anxiety can collectively build up in organizations. When this occurs, people need a way to deal with the heaviness. Most of the time, people cope with the anxiety by trying to dilute it. So we dilute it to make it less noticeable. There are a number of tools that people in organizations use to dilute anxiety. The problem with these tools is that they usually work for the short-term—they do indeed make the anxiety less noticeable. However, the tools also cement the anxiety and make the anxiety permanent. These tools "institutionalize" the anxiety. This process actually increases the anxiety over the long-term!

The interactions and patterns that develop in response to the anxiety often become permanent in the system. So the anxiety remains in the system regardless of the individuals in the system. The anxiety gets passed around, and passed from generation to generation within the system. This can be easily seen in comments that reflect a nagging fear of what might occur.

There are many tools to dilute anxiety, but here are four common methods.

Anxiety is different than stress. Stress is more like a hammer on the head. Anxiety is more like a hand around the throat.

- The first tool to dilute anxiety is the use of triangles. When two people have a conflict, they triangle in a third person (see the section below on *emotional triangles*). As we mentioned, this process makes the anxiety less noticeable in the short-term, but prolongs it and increases it in the long-term.
- Another tool for coping with anxiety is to project the anxiety. Projection is when we label the anxiety and then accuse someone else of the anxiety. Projection is the Macbethian idea, "methinks he doth protest too much." It's when the television evangelist proclaims tirelessly about the evils of sexual immorality, only to find he is having secret rendezvous with prostitutes. Projection is when the man who secretly struggles with his own sexual orientation, but plays the role of the playboy, publicly trying to show himself as a macho ladies man (he is trying to convince himself and others that he is a virile heterosexual). When we project the anxiety, then we can think that it is someone else's problem, not ours.
- Another tool for diluting anxiety is to spread it around. This is when we are conflicted in one relationship, so we spread the anxiety around by conflicting another set of relationships. For most of us, a variety of low-level conflicts are better than one high-level conflict. Consequently, we can create conflict simply to dilute the other conflict! You may have had the experience of someone "picking a fight" with you for no apparent reason. It

may be that they were trying to use a conflict with you to dilute their own anxiety.

- A fourth tool for diluting anxiety is to attach it to something. We are full of anxiety, but we blame the feelings we have on some insignificant detail. Whenever there is an overfocus on a particular detail, chances are that it is related to anxiety. Whenever we see people who become consumed with a tiny detail, it is possible that the real issue is anxiety—and the person has simply attached the anxiety onto an insignificant issue.

Differentiation

Differentiation is one of the pillars of the systems theory articulated by Murray Bowen (1978). Differentiation is walking the line between "being autonomous" and "being in relationships." It has to do with the perception of personal boundaries. It is the emotional maturity of knowing the difference between "my issue" and "your issue." Differentiation is the ability to articulate your own goals and still remain connected to those people who disagree with your goals.

> *Differentiation is an essential concept in Process Adaptive Systems. It might even be the most important concept.*

Differentiation can be understood in terms of a spectrum, with *completely undifferentiated* on one end, and *completely well differentiated* on the other. Probably no one reaches either extreme.

Normal, healthy living in communities calls for people to be intimate with some people and autonomous with others. Sometimes we need to be our own person, express our own personality, and do our own thing. Other times, we must be part of a group, and be in relationships with others. Differentiation is walking that line—it is regulating the line between closeness and distance. A person who cannot be intimate with anyone is poorly differentiated. A person who is intimate with everyone is also poorly differentiated.

Differentiation is to articulate your own goals, even in the presence of others' opposite or competing goals. But differentiation is always in the context of a relationship. Anyone can articulate their own goals and then say, "to hell with you if you don't like it." A well-differentiated person can articulate their own goals but still remain connected to those with opposite goals.

Well-differentiated people have good, healthy boundaries:

- They can control their emotions. In decision-making, they can separate the intellectual and the emotional. They still have emotions, of course, but they are not *controlled* by their emotions.
- They know where they stand on a given issue, and can live with others who don't agree with them. They are less influenced by either praise or criticism.
- They can be autonomous, even when the others around are saying, "we should all believe the same thing."
- They can enter relationships and not have life governed by the relationships. A well-differentiated person will make a conscious choice to have a life based on intentional principles and goals, and not be ruled by emotionally reactivity.

Poorly differentiated people have difficulty defining their boundaries.

- They operate with high emotional reactivity. They react emotionally to everything. Life, for these people, is about feeling good or feeling bad.
- They have difficulty taking a stand, because when another person states a belief, they feel they must conform.
- They are chameleon-like, taking the belief systems of whoever is around them.
- They have trouble keeping personal information; they share private information about themselves too easily; they can be invasive and nosy; they can pry into other's affairs because they have few boundaries.
- They have a poor ability to reason out solutions, and react quickly and emotionally to almost every issue.
- They have less of a repertoire in dealing with stress. They have few options when they are stressed or threatened except to strike out emotionally. Because of this, families that are poorly differentiated tend to become more symptomatic (they end up in therapy more often).

People can be poorly differentiated or well differentiated. But this is true for whole systems, too. An entire system can be described as poorly differentiated or well differentiated. Organizations can be invasive and nosy, just like individuals. Organizations can have a pattern of emotional over-reaction, just like individuals. Organizations can discourage people from taking a stand, in the same way individuals can shy away from autonomy.

The level of anxiety in a system has a big role in pushing the level of differentiation one way or another. As anxiety increases, the system is pushed toward poor differentiation. When anxiety increases, so does the tendency to react emotionally—and to lose sight of the big picture. Anxiety puts pressure on the system to become less differentiated. Lower levels of differentiation increase anxiety. It can become a vicious downward spiral.

Organizations that are poorly differentiated tend to become mediocre, because members focus on "not hurting anyone's feelings." They are more concerned about turf, power, and "not rocking the boat," and less concerned with excellence and carrying out the mission of the organization. In some poorly differentiated organizations, conflict avoidance becomes a frequent way of dealing with conflict. They decide on whether a course of action is good or bad based on who will be the winner and loser (rather than deciding if a course of action is good or bad based on if it meets the goals of the organization).

Emotional Triangles

A triangle, you may recall, is a way that we dilute the oppressive feeling of anxiety.

Triangles happen like this: Two people can't get along with each other. They fight, bicker, and spend long periods of time not talking to each other. Their fighting is uncomfortable because of its clashing and unpredictability. It produces anxiety in both partners, because people tend to dislike conflicted relationships. So, the two fighting partners pull in a third person as a "go-between." Both fighters can now relate to this third person. This stabilizes the relationship, making it more comfortable for the fighters (but more unpleasant for the person who was triangled in!). Initially, the anxiety is diluted— the anxiety is made less noticeable. However, in the long run, the conflict is cemented, made permanent, increased, and the anxiety simply shifts toward the new person who was triangled.

Figure 2
Differentiated People

Poorly Differentiated	Well Differentiated
Has difficulty expressing own goals when the listeners will not agree	Autonomous; expressing own goals
Difficulty knowing when intimacy is appropriate. Can be over-intimate or never intimate.	Able to be psychologically intimate with others when appropriate.
Quickly offended, over-sensitive.	Manages emotional self
Highly reactive; reacts quickly and emotionally to any situation	Responds intentionally in a non-anxious way
Intolerant; thinks in black-white and either-or	Takes the broad picture; knows that dichotomies are false; understands shades of gray
Quick to blame and criticize; has difficulty taking responsibility—it's always someone else's fault	Takes responsibility for own actions
Constantly on the defensive	Emotionally relaxed
Tends to be vague, underhanded, covert	Tends to be clear, objective, and willing to be open
Sees life in terms of "who is for me and who is against me"	Able to participate in life without desiring that they be the center of life
Thinks everything is about me; believes they are responsible for others	Believes they are responsible for their own self

- A married couple doesn't particularly enjoy each other's company. So, they continue to have children so they won't have to deal with the anxiety in their marriage. The parents have triangled in their own children to stabilize the conflict and dilute the anxiety.
- In a corporation, two departments can't get along with each other. So they hire a person to be the liaison. This kind of thing is rarely successful, and the anxiety will probably find it's way to the liaison, and this person will probably burn out. Hiring this liaison is a sure way to make the conflict permanent.
- In a church, the pastor doesn't know how to manage a large congregation, and the congregation knows the pastor is a poor leader. So, they manufacture a budget deficit, so that when they come together, they could discuss the budget, rather than discuss the fact that the pastor is an ineffective leader. Discussing the budget was more comfortable than discussing the pastor's abilities. Notice here, that it is not necessarily a person who becomes triangled. In this case, the two parties triangled in a thing—a budget deficit that they manufactured. This manufacturing process was beneath the level of awareness—no one knew that they were manufacturing a budget problem.

When two people are calm and comfortable, they don't need to triangle anyone into their relationship. However, when the anxiety between two people grows, the "natural" thing is to triangle in a person, place, or thing. This is such a natural process that people rarely realize that it is happening. We frequently create triangles and allow ourselves to be drawn into triangles without even knowing it!

Triangles interlock, too. Figuring out interlocking triangles is double the fun.

When anxiety is high, and differentiation is low, triangles are widespread in the system. This proliferation of triangles, in the long run, increases anxiety and continues to push the level of differentiation lower. Triangles are not only a symptom of poor differentiation, but they are also a factor in pushing for decreasing differentiation.

The proliferation of triangles and their destructive effects can be stopped, of course, when individuals refuse to be a part of a triangle. This means being aware of triangles and setting up boundaries. It

Figure 3
How Triangles Form

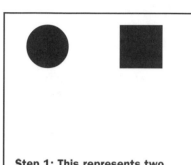

Step 1: This represents two people.

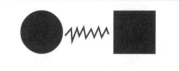

Step 2: The two people have a conflict.

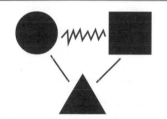

Step 3: The two conflicted people triangle in a third person who stabilizes the conflict and temporarily dilutes the anxiety.

Step 4: The anxiety of the conflict is transmitted to the third person, who feels the anxiety and the heavy weight of the conflict.

means not being drawn into a triangle. It means no longer rescuing and perpetrating the anxiety of the other two. It means no longer playing the game. However, when a person detriangles, the other two will muster immense pressure to "get back in the game."

Forces of Togetherness

Edwin Friedman (1999) coined the term *forces of togetherness* to name the push toward thinking the same. The forces of togetherness push us to be alike in thoughts, feelings, and behaviors.

When anxiety is low, differentiation and autonomy are high. On the other hand, when anxiety is high, or when the organization is full of poorly differentiated people, then the *forces of togetherness* gather steam. There is pressure to think the same. Autonomy is punished. Loss of individuality is praised. People in the system are treated as if they are responsible for the happiness of others. Poorly differentiated people experience a decrease in their perceived level of anxiety when everyone is together in thought, word, and deed. So, when anxiety is high and differentiation is low, you can expect the forces of togetherness to push you away from your own goals and activities, and toward the goals and activities of the group.

Imagine for a moment, ten adults stuffed into a phone booth. Each time one person moves, everyone else feels the movement. If one person raised an arm, someone else would be elbowed in the face, causing that person's head to knock another head. Every time someone tried to reposition, everyone else would feel that movement. There would be a counterbalance and interplay between the forces of togetherness and individuality. In a system, when differentiation decreases, individuality is less developed and togetherness needs are stronger. The togetherness needs of a poorly differentiated person are deep yearnings to be loved, accepted, and guided through life. When the differentiation of the system increases, individuality is better developed, and the forces of togetherness push and pull with less intensity. In a poorly differentiated system, there are ten adults stuffed in the phone booth; in a well-differentiated system, there are two adults in the phone booth (not even uncomfortable, depending on who is in there with you).

If you want a video on systems, go rent **Mindwalks**. *In it, a burned-out politician, a second-rate poet, and an exiled physicist talk about systems. It's not exactly car chases and gunfights, but it's a good discussion of complexity.*

Perhaps organizations should hold strategic planning meetings in phone booths, to match the emotional climate they have created.

The forces of togetherness create fusion. No one is allowed to have independent thoughts. Boundaries become more and more permeable.

Frequently, poorly differentiated leaders put positive spin on the forces of togetherness. It is cast in the rubric of "getting on the same page," "all speaking the same language," or "wasn't it great ten years ago in our organization when we all trusted each other." Despite the euphemisms, the forces of togetherness are a negative dynamic in the system. The presence of togetherness forces is a symptom of chronic anxiety and poor differentiation.

In a poorly differentiated family system, the forces of togetherness expend considerable energy keeping the system together. Regardless of the stage of the family life cycle, they will have "stay together" as their primary value. But it is not a "stay together" that respects individual freedom and independent thought. It is "stay together" precisely at the *expense* of freedom and thought. In a poorly differentiated family system, there is a stay together mentality, no matter what the individuals feel like; for these families, "feelings are unimportant; it's how things *look* that matters." Sometimes, the family members have to engage in a full-scale act of rebellion to get out of the system that has become a prison—so they have an affair, dye their hair green, run away, get pregnant, or join the circus.

In an organization, the forces of togetherness will put a drag on a differentiated leader. When the forces of togetherness rule, the followers will resist any efforts by the leader to change the system. This is why incremental organizational change is rarely effective when system anxiety is high.

The Identified Patient

When you diagnose the problems in a system, *the identified patient* is where the symptoms erupt. The identified patient is not where the problem is—it is where the problem shows up. The identified patient is the proverbial scapegoat.

- If you get an eye infection, the problem is bacteria in your eye. What shows up is a red eye. You can get the red out by using Visine™; however, Visine™ does not kill bacteria—it only makes your eye less red. Using Visine™ for an eye infection will make the problem (the bacterial infection) worse. The red eye is the identified patient—it is not the problem, but it is where the problem shows up.

- Two parents bring a youth to a therapist. The youth has green hair, studded wristbands, a wide assortment of facial jewelry, and leather clothes. The parents demand that the therapist "fix" their child. Chances are, it is the parents, not the youth, who has the problem. The anxiety of their marriage relationship had become intolerable, so they triangled in their child. They, in a sense, elected their child to be the problem. They manufactured their child's rebelliousness so that they wouldn't have to deal with their anxiety. The youth is the identified patient, the scapegoat. However, the youth is the symptom of the problem, not the problem. If the therapist treats the symptom, it will cement the youth's rebelliousness and the parents' lack of responsibility. If the well-meaning therapist "cures" the youth, the parents will simply transfer their anxiety to another person or thing.

- In a highly conflicted organization, the poorly differentiated middle manager may be the first person to melt down. The middle manager may feel the stress and anxiety of the organization, and so he has temper tantrums at meetings, makes illogical decisions, and whines incessantly about the "good old days." It's easy to blame the middle manager, saying, "He's the problem." However, it may be that this manager was the identified patient—he is only the place where the symptoms have erupted. The problem may be much broader—conflict and anxiety in the system. However, like a canary in a coal mine, the poorly differentiated people will feel the pressure most saliently. If the middle manager is fired, the symptoms will simply erupt somewhere else.

So, the problems you see in your organization are probably systemic problems. If you attack the symptom, the most readily apparent problem, the problem will get worse. You may fix that particular symptom, and everyone will give you positive strokes for your good leadership, but the problem will remain, and just shift into another symptom.

Chapter 7

Some Basic Principles of Complex Systems

1: *Systems are interdependent, so what you do will have intended effects and unintended consequences.*

There will always be consequences of changes and actions. Some will be intentional, others will not. Senge (1990) suggests that most of today's problems come from yesterday's solutions. Careful leaders will think through possible scenarios of unintended consequences. Further, consequences will often show up in the most vulnerable part of the system.

2: *Systems thinking is frequently counterintuitive. You can make the changes you want, but usually not the place where you think.*

In complex systems, if a leader wants effective, fundamental change, the leader must not implement strategies at the point where the problems erupt. Fixing the identified patient will simply change the visible symptom—but it won't fix the problem. Rather, leaders must make changes at the point of greatest leverage. Otherwise, changes will not last, or the changes will only alter the symptom and not the problem.

3: *Persons and organizations find ways to cope with anxiety.*

When anxiety is high, people will find a variety of ways to cope: Triangles will increase. Secrets will become more salient. The forces of togetherness will gather steam. Identified patients, or scapegoats, will be blamed. Complaining and fussing will increase. While none of these activities will fundamentally decrease the anxiety, these activities do make the perception of anxiety less oppressive for individuals. But the cost is that the anxiety is maintained, and it increases in the system. Anxiety begets anxiety.

4: *Organizational problems are rarely neat, clean, easily diagnosable, and solvable with a simple solution.*

Organizations have problems, and every problem has multiple symptoms. It is important to find the underlying problem.

The problem, of course, is that symptoms are easy to see; underlying problems are not.

5: *If you see a solution to a problem that's too easy, it probably is.*

Linear thinkers see easy solutions to complex problems. And frequently, the solution will make some immediate, short-term positive impact, thereby reinforcing the linear approach to problem solving. But there are frequently long-term disastrous results to short-term problem solving. For example, a few years ago, many businesses saw shrinking corporate profits. Their linear thinking solution was to lay off hundreds of employees. The corporations saw immediate benefits—they would finish their fiscal year in the black. However, research has suggested that most of these corporations had continuing problems after the layoffs. Low morale and increased anxiety created lower quality production, which drove away customers, causing more layoffs and increased anxiety, which further lowered quality, driving away even more customers. Complex problems usually have complex solutions.

6: *The two most important keys to understanding how an organization works are to understand the levels of homeostasis and differentiation.*

The stated goals of the organization, the mission statements, and the vision of the institution, are all important. But you will know much more about the organization when you know if it is well-differentiated or poorly differentiated. This will tell you if the organization is capable of change and growth. If you know its level of homeostasis, and the forces that push for homeostasis, then you will know how the organization maintains its level of balance—and what you should expect in making a change.

7: *Organizations have common, predictable patterns of behavior that can be understood from a systems perspective.*

Whether the system is a human services organization, a corporation, a community, a military installation, or a family, certain patterns are repeated. There are common patterns of behavior that need a systemic perspective and a systemic intervention.

Figure 4
How to Differentiate in Ten (not so) Easy Steps

In an undifferentiated environment, boundaries are too permeable, invasiveness abounds, and pathogens are dysfunctioning all over the system. The forces of togetherness are strong, and attempts by individuals to differentiate are met with terrible amounts of pressure to "stop that independent thinking and quit having boundaries." It is even more difficult when the leadership is creating and or condoning these dynamics.

Here's a bunch of suggestions. All of them are easy to say and difficult to do.

1. Self-define always. Just think about what you are, who you are, what you will do, and what you won't do.

2. Don't get sucked into the whirling vortex of the forces of togetherness. Be aware of triangles and don't get drawn into them.

3. Realize that what the organization does is not a reflection on you or who you are. Separate yourself—emotionally speaking—from the organization.

4. Get a good support system outside of work. You'll need it.

5. Resist the temptation to define others. Don't blame others. Don't tell others what they think. Don't pretend to be a mind-reader.

6. Don't try to change anyone else.

7. Don't tie your joy, your happiness, or your satisfaction to what other people do. It's dangerous to give other people power over your own happiness. Don't give them that power. Tie your job satisfaction to what you completely control.

8. Make clear boundaries. When in doubt, err on the side of impermeable boundaries. Toxic forces and pathogens hate boundaries. A boundary is like garlic to a vampire.

9. Allow other people—especially the leadership—to do whatever they need to do. You can't control them. You just do what you need to do.

10. Consistently, constantly, and with great intentionality, self-define.

Chapter 8

Why Do We Need Systems Thinking, Revisited

E ach of the common problems below can be informed by a systemic principle or key concept:

- *A reoccurring problem that won't go away, no matter what you do to solve the problem.*
  Usually, this type of pattern is because well-meaning leaders see the symptom of a problem, and they try to solve the symptom rather than the underlying problem. In fact, usually, when you "cure" a symptom, you make the problem worse.
- *Lots of complaining and sour attitudes; from month to month the complaining continues, just shifting the topic of the complaint; there is even complaining about the complaining.*

Rampant complaining is usually a sign of organizational anxiety. When anxiety is high, complaining is rampant. Sometimes, administrators will try to "fix" whatever people are complaining about. However, that will not stop the complaining; the complaining will simply change topics, because the complaining reflects the systemic anxiety. Addressing the content of the complaining will not address what is underlying the complaining (the processes that drive the anxiety).

Here's one of the ways that you can know if you're "getting" systems thinking: try to explain it to someone else. If they tilt their head and give you the "confused dog" look, you can rest assured you're talking systems. Or maybe you're just obtuse.

Anxiety is not due to events that occur in the life of the organization. Anxiety grows from a family of origin, and is made worse by long processes of decreasing differentiation, organizational secrets, and growing lack of trust in the leadership.

- *Organizational secrets—things that you think go on but no one talks about.*
  Rumor mills are signs that organizational secrets are active. Secrets are not about "who makes what salary," or "what punishment so-and-so got for their mistake." Secrets are the things that *should* be talked about, but someone fears that "if this gets out, it will shame the institution, and therefore me." Secrets may be about opinions, values, or scenarios when those things are not permitted to be spoken out loud. Secrets will perpetuate and increase anxiety.

- *Policies and rules that actually hinder the stated goals and purposes of the organization.*
  Some organizations thrive on developing mission plans and purpose statements. These plans and statements can be powerful tools to move an organization. However, if there is no attention to the culture and the systemic nature of the organization, there will be no movement toward the stated goals. In fact, it is natural for a system to put policies and procedures into place to maintain a homeostasis—even if that homeostasis is antithetical to the purposes and goals.

- *Organizations that never seem to get anywhere; they just stay busy fighting the daily fires.*
  Some organizations thrive on crisis. They like crisis, and if there is not a crisis, they will create one. This is most common in poorly differentiated leadership, who will have knee-jerk reactions to every problem. They will be more interested in putting out fires than preventing fires. Poorly differentiated leaders are less concerned with the vision, and more concerned with "why can't we all just get along?" Many of these leaders will create a peace-at-any-cost mentality; Friedman (1999) calls these kinds of leaders "peace-mongers." By their nature, they will increase the use of secrets, create anxiety, and embolden the forces of togetherness.

- *Logical people who sit in committees and make illogical decisions.*
  The level of differentiation, the secrets present in the system, and the amount of anxiety in the system have much more to do with

the decisions than any rational process. Rationality and logic are extremely weak influences compared to the power of poor differentiation, high anxiety, and secrets. Especially in high-stress times, people will tend to think more linearly, more reductionistically, and more mechanistically.

- *Organizations that never seem to get anything done.*
  Homeostasis is a powerful force in an organization. "Keeping things the way they have always been" is rarely stated but frequently valued. When forces of togetherness are high, it is difficult to get new things done. People will complain about the stagnant nature of the organization, and everyone will have ideas about how to change it. But, usually beneath the level of awareness, people hold the value of "a known stagnation is preferable to an unknown future."

- *A few people, usually the fussiest and most-disliked, seem to control the agenda of the organization.*
  In an organization that is poorly differentiated, people don't like to make anyone feel bad. So, it becomes easier to act on the whinings of the fusspots, rather than make substantive progress or change. Any decision makes people fuss; and when organizations are afraid of fussing, they will spend their time trying to answer whining than risking a decision. The organization may frame the issue as "we're being sensitive to people's expressed needs." But frequently, it is simply that that they are afraid of making decisions that will surely increase the complaining. So, in an attempt to "fix the fussing," they actually prolong the fussing and make it worse in the long-term.

- *The harder the leader works, the less that seems to get done.*
  When one part of a system overworks, another part will underwork. This maintains a balance in the system. This will forever confound linear-thinking leaders, until they realize that the level of homeostasis must be raised to a new level.

- *The more an organization tries to change, the more they seem to stay the same.*
  Homeostatic influences push for sameness. The push is even greater when anxiety is high. The push for sameness is even greater when leaders are poorly differentiated. Change is sabotaged by a variety of forces—secrets, triangles, created

crises, personal attacks on the leader, blaming, shaming, etc. Change is possible in organizations, but rarely do you push where you think it needs to change.

- *A lot of people expressing a desire to become a learning organization, but no one knows how to become one.*
  The learning organization is an important concept. In a rapidly changing environment, it is critical that organizations can learn about their experience and their environment and make the necessary changes. However, as Senge (1994) points out, it is easy to learn when the feedback is clear and immediate (this is how we learn to walk); it is extremely difficult to learn when the feedback is ambiguous and delayed. It takes people who are very perceptive and very attentive to the learning process; it takes people who can work to change the anti-learning forces in the organization. And it takes well-differentiated, non-anxious leaders who will maintain their course when (not "if") the sabotage to change comes.

- *Whenever the organization tries to change, a crisis happens that stops the change because people have to deal with the crisis.*
  There are few coincidences in organizations. Change is sabotaged at a variety of levels. Frequently, people are elected to have crises in order to move the energy and resources of the institution to deal with the crises rather than work for substantive change.

Chapter 9

Summary

S ystems thinking is easy for some, difficult for others. Some people intuitively think in systemic terms. Some people have thought systemically all their lives but have never labeled it as "systems." However, most people today think in linear, reductionistic, and mechanistic terms.

When people start thinking with systems, it appears to be a little chaotic. There appears to be too many variables and people think, "how can I do anything if I don't know what effect my intervention will have?" This kind of thinking is normal at first, and usually gives way as various principles of systems behavior become clearer.

Systems thinking brings powerful tools and enlightened perspectives to organizational diagnosis, problem-solving, strategy, and leadership.

Free Lawn Tips

People with grassy lawns usually like their lawns healthy, green, and weed-free. Healthy lawns are a result of several factors: enough light; enough nitrogen and other nutrients; enough water; organic matter in the soil; lots of worms that aerate the soil (the worms make holes through the ground, which loosens the soil so that roots can grow, and allows air to get to the roots, and the worms deposit further organic matter into the soil).

However, people sometimes mistake *green* lawns for *healthy* lawns (a classic linear mistake). So, many people buy chemical fertilizers. This will make the grass green (or it will kill the lawn if you put too much on). One of the problems, of course, with chemical fertilizers is pollution (a yard cannot use all the nitrogen, so it runs off into the ground water, into the rivers, making the algae bloom, which use up all the oxygen, causing fish to die or leave, which causes the river to become stagnant). But another problem with the chemical fertilizers, particularly the ones with weed killer in them, is that worms and microbes don't like chemicals. Worms leave, which mean less organic matter in the soil, and less aeration. Less aeration means more compact soil, which further hinders wormic activity (actually, we're not sure if *wormic* is a word, but it sounds cool); which means less organic matter in the soil, which further compacts the soil. Compact soil will not hold water well, which dries out the soil, further reducing biological activity. So, over the course of several years, heavy use of chemical fertilizers and weed killers can create an unhealthy lawn—because compact soil, lack of water, lack of biological activity is a poor place for grass to grow, but a great environment if you're a fungus, parasite, or weed. So, the grass will be weak, fragile, and unhealthy, *but it will be green*!

Every lawn enthusiast wants healthy grass. The linear thinker will work to make the grass green. The systems thinker will work to make the soil healthy. When the soil is healthy, the grass will become green on its own.

Every good lawn needs a good systems thinker to care for it.

Part Two: There are No Things: Linear and Systems Thinking

Chapter 10

Newton's World

I saac Newton (1642-1727) created a new world. Prior to Newton, the defining way of organizing reality was religion. Religion (in western Europe, it was primarily Christianity) served as the central metaphor, the creative focus, and the ultimate basis of legitimation. So, when people asked, "why does it rain?" the answer was, "because God made it rain." When people asked, "Why do the sun and planets move around the earth?" the answer was, "because God said so." When people assumed (incorrectly) that the planets moved in perfect circles, the rationale was, "God would never create anything imperfect."

At age 12, Isaac Newton started to build things and experiment in a laboratory to impress a girl. Eventually, she married someone else. Isaac never married.

But Newton changed all that (actually, Newton was the most salient of a long line of adventurers; Newton put together the thinking of those who had gone before him).

Ian Marshall and Danah Zohar (1997) recount the "philosophy" that Newton created. Newton said that the universe was based on static principles that could be understood with proper observation. In other words, if we study it enough, we will understand why it rains, why the sun shines, and why gravity holds us to the earth. For Newton, we lived in an orderly universe, and there were a few basic principles that could be used to explain everything. Simply by reducing everything to the component parts (the philosophy of reductionism), we will discover the rigid laws of cause and effect (the philosophy of determinism). Newton's universe was a vast machine, and this machine metaphor became the central focus that colored our speech and our thinking (Marshall, 1997).

For Isaac Newton, and the philosophy that followed him, the world was predictable, reducible into a few principles, and every effect had a cause. This "machine thinking" affected our understanding of everything. We saw every aspect of reality as orderly cogs in a giant machine.

In physics, the argument that began with the ancient Greeks was finally settled. All matter was made up of molecules, and all molecules were made up of atoms. Atoms are always the same; they don't change. They were the fundamental building block of matter. Later, we came to believe that atoms consisted of three particles—protons and neutrons that lived in the center, and electrons that circled the outside. These particles were the basic, elementary building blocks of matter. These particles were irreducible—they could not be broken down any further (or so people believed).

Everything was a machine—orderly, predictable, and available for discovery. All things could be studied by a detached observer.

Fredrick Taylor, the first organizational theorist, described organizations as machines. We still talk about organizations in terms of "cogs" and "hard wired" and "squeaky wheels." If the organization works well, we call it a "well-oiled machine." This philosophy led to the idea that employees are interchangeable, and should simply do the job they are given. Various methods can be used to increase their efficiency when they do their job. Punishments can be doled out when an employee is performing poorly—and we have believed those punishments will have no consequences. As if.

> *Taylorism, the management style that views organizations as machines and employees as cogs, is still the dominant way people see organizations.*

For maximum efficiency, the organization needed three things: 1) one person at the top who knew everything; 2) a herd of mindless automatons who would work on the assembly line; and 3) a few middle-level managers who would poke the automatons with sticks when they moved too slowly. This kind of organization was neat, clean, and efficient. And, in Newton's universe, this kind of organizational hierarchy made sense. After all, the organization was a machine, and the employees were mindless automatons who themselves were expected to work like machines.

The machine metaphor became the central metaphor, the creative focus, and the ultimate basis of legitimation. And all was well with the world.

And then along came Albert Einstein.

Chapter 11

Einstein's World

Albert Einstein (1879-1955) created a new world. He left Newton's machine world to create a quantum world (actually, Einstein was the most salient in a long list of cognitive adventurers).

In quantum world, electrons, protons, and neutrons are not the smallest bits of matter. Each of them can be broken down into quarks or leptons. Quarks, for example, come in different kinds of electrical interactions, and are classified in terms of charms and flavors. The idea that protons and electrons were the irreducible was blown apart. When scientists delved into the smallest bits of matter, they realized that the smallest particles are no longer particles—they are only interactions.

They actually saved Einstein's brain in a jar. Scientists periodically run tests on it to find out why it was such a good brain.

In Einstein's world, when we get to the smallest bits, we find that *there are no things*, only interactions.

In a Newtonian world, matter is matter, and two bits of matter cannot occupy the same space. But today, we know that certain particles, like electrons and photons, can indeed occupy the same space. When they occupy the same space, they begin to behave with the properties of a field—they begin to exert force on the particles around the field, in the same way gravity exerts force on objects within its field.

In a quantum universe, you have black holes and wormholes. Black holes are collapsed stars that keep collapsing until they are so dense that the gravity becomes immeasurable. Nothing, not even light, can escape their gravity. Einstein and Hawking have suggested that the black holes keep collapsing until they wink out of existence. Then, they form a wormhole. Going into a wormhole would send you to another place in the universe, or perhaps another time…

Or perhaps another universe… Or else you'd just be crushed by the gravity of the black hole.

In Newton's world, time was constant. Time, in fact, was the ultimate in the machine-metaphored world. Time went at the same rate all the time. The closest thing we have to relativity-of-time in Newton's world is daylight savings time.

However, in Einstein's world, time is relative. The rate of time depends on velocity. If you sent a spaceship to the nearest star and back, and the spaceship was traveling 186,000 miles per second, it would travel for a little over eight years. So, upon their return, the people on the spaceship would be eight years older. But on earth, thousands of years would have come and gone.

Today, scientists are doing things that, a few decades ago, would have been science fiction. Only a few years ago, the speed of light was considered immutable. However, scientists recently slowed light down to a crawl. Further, scientists recently (Zabarenko, 2000) coerced light to go faster than the speed of light. Consequently, light exited the experimental tube before it entered the tube (yes, you read that correctly).

Subatomic particles don't follow normal (read: *Newtonian*) laws of physics. They operate with quantum laws. The textbook metaphor for a quantum world is Schrödinger's cat. Erwin Schrödinger invented (discovered?) this metaphor. In this metaphor, we have a cat in a box. The box has a nefarious device in it: there are two levers. One lever, when pressed, will give a piece of cat food; the other, when pressed, will give poison. In Newton's world, the cat would press *either* the food lever, or the poison lever. The cat would get *either* food or poison; and then the cat would *either* be dead or alive. Opening the box and looking at the cat would not change whether it was dead or alive. However, in Einstein's world, all possibilities co-exist constantly. The cat eats both the food and poison; the cat is both dead and alive at the same time. Only when you open the box and look at it does it become dead or alive; our looking has either saved the cat or killed it.

The 17th century French philosopher Rene Descartes believed all sciences stemmed from physics. We believe biology is the central science.

We review some of the strangeness in theoretical physics these days not to try to understand it. Really smart people have been working for years to understand this kind of thing. So, there could be at least five points here for the rest of us.

Point #1: Systems thinking is a new way of thinking. The field of physics had to move from a Newtonian view to a quantum view. The field of physics is still adjusting to a new way of understanding the world. And, when we move from a linear view of the world to a systems view of the world, it is similarly strange.

Point #2: The traditional way we have thought about organizations has stemmed from Newton's world. We don't want to discard Newton's world, we simply want to recognize its limitations. We also need to know that the traditional (read: *Newtonian*) way of understanding organizations borrows heavily from linear thinking. When we move to an Einsteinian view of organizations, it seems strange at first. It seems chaotic, unpredictable, and messy.

> *With all due respect to my high school physics teacher (who gave me a "D"): Physics is dead. Long live biology.*

Point #3: When you get down to the smallest bits of matter, those particles are no longer things—they are interactions. Similarly, in organizations, the most important part of understanding how things work is the interactions. We can read mission statements, job descriptions, and organizational flowcharts until we are blue... but if we don't understand the patterns of interaction between people, we will not understand how the organization works.

Point #4: We know that when subatomic particles occupy the same space, they take on properties that are not present with individual particles. The properties they take on—the field—exert influences on the neighboring particles. Fields, like gravity, exert influence. Systems always exert influences on itself and the neighboring stuff. Individuals don't always exert influences—but systems always do!

Point #5: Organizations today, like Schrödinger's cats, are unpredictable, nonlinear, non-reductionistic, where all possibilities are open at once. Organizations today are in a chaotic, messy muddle of opportunities and dangers. We like to believe that we can control organizations, lead them, and implement change whenever we want. We are comforted by the traditional hierarchical model of organizations, because we grew up with it, have lived with it, and we understand it. However, it is an incomplete way to understand the system. Since today's organizations are complex systems, we need a systemic perspective. If our organization is in an environment that is predictable and non-complex, then linear thinking will work just fine (although, I don't know of any organization that is in that kind of environment). However, a systems view gives impressive insight into organizations that are unpredictable, nonlinear, and full of endless possibilities.

Chapter 12

Linear vs. Systemic Thinking

Linear thinking is cause-effect thinking. One effect has one cause. Sometimes linear thinking works adequately. If you are driving down the road, and you run out of gas, your car will stop. Your car stopped (effect) because you had no gas (cause). If you put in gas again, your car will go again. Linear thinking is very adequate when your car runs out of gas.

However, human relationships are never as easy as that.

For firefighters, linear thinking is important. Firefighters must put out the fire. There's a fire, put out fire… that's linear thinking. That's perfectly adequate. If my house was on fire, I wouldn't want a firefighter to stop and ponder about multiple effects and recursive influences of the fire's origin. A firefighter who stops to ponder all that stuff will be fired from the job (no pun intended).

First century Rome employed 7,000 fire fighters.

However, that's not all that firefighters do. If firefighters *only* had to put out fires, linear thinking would be okay. But they also need to consider (preferably after the fire has been extinguished) why the fire started. They need to consider the origins, the environments, and the way buildings are structured. They have to tell people, "No, you can't build your house that way, because of the fire danger." They need to know why fires start, the different kinds of fires, and how to extinguish each type of fire. Most importantly, they need to make sure people build their houses and businesses in a way that minimizes the possibility of fire. If firefighters didn't do this work, fires would be commonplace. But they use systemic thinking, and so they are able to prevent fire. If they didn't use systemic thinking, they would spend every waking moment putting out fires.

If fact, many organizational managers are like linear firefighters. They never think systemically, and so never understand why the fires keep happening. Consequently, in a good linear fashion, they spend all their time handling crises. Ironically, we call this kind of work, "putting out fires." They are like the firefighter who never worked to prevent fires.

Systems thinking realizes that most of the time, a variety of components have a variety of influences on the effects, and the effects cause changes in the original components, thereby influencing the influences, and changing the components, the influences, and the effects.

> *We frame systems thinking against linear thinking.*

The problem with linear thinking is that in relationships and organizations, the apparent effect rarely has a single cause. Usually, it doesn't even have multiple causes. In human interaction, effects are usually the result of many effects and components that influence each other.

In cause and effect (linear) thinking, there is a belief that a problem can always be traced back to a cause—usually a single cause. Linear thinkers will try to find the cause, eliminate that cause, and then believe that the problem will magically go away.

- A few decades ago, mosquitoes were a problem. So, linear thinkers came up with a solution: a pesticide called DDT. Proponents of DDT told people it would kill mosquitoes, but it was safe for plants and animals. Proponents of DDT would actually eat a teaspoon of DDT each day to prove it was safe (DDTea?). The problem of course, is that the effects of it did not show up for several years. Birds were born with soft eggshells. The brush with DDT almost wiped out several species of birds. Today, decades after the cessation of DDT, it is showing up again in women's breast milk—and no one knows if that's harmful or not. As an ironic footnote, within a few years mosquito physiology had adapted so that they were immune to DDT, and with the death of the mosquito-eating birds, the mosquito population swelled. The linear solution (pesticide) made the problem (mosquitoes) worse.
- A conflicted family goes to a therapist. Mom and dad are having a problem with Johnny, the teenage son. He is mouthy, talks back,

and is generally insensitive and verbally abusive to the other family members. He has green hair, facial jewelry, and studded wristbands. A linear thinking therapist might work with Johnny, try to teach him social skills and communication skills. On the other hand, a systemic thinking therapist would first wonder why Johnny has been elected to be the problem. The systemic therapist might suspect that the parents can't get along, and so rather than deal with their problems, they unconsciously pass the problem on to the teenage son. The linear therapist would cement the anxiety between mom and dad, even if Johnny were "cured."

System (sys-tem) noun. a regularly interdependent or interactive group of items forming a unified whole; from the late Latin systemat, meaning to combine.

• Many politicians believe that "the family is dying" (a dubious proposition. Every generation in human history has believed the family is dying, and it is still as strong and weak as ever). The next step in some politicians' dubious logic is "the family is dying because there is so much divorce." The logic continues: "Divorce bad," and "divorce is too easy." So, the final conclusion becomes: "if we make divorce more difficult, that will make the family better." As dubious a proposition as it is, this kind of legislation is appearing in most states. However, about twenty years ago, it was extremely difficult to get a divorce. Then, the laws changed (no-fault divorce) and it became much easier to get a divorce. However, *divorce rates before the law and after the law were the same.* But, when divorce became easier, reports of spouse and child abuse dropped sixty percent.

Linear (linnee er) adjective. developed sequentially from the obvious without indepth understanding; from the Latin linearis, meaning line.

• In an organization that makes widgets, there is a lot of complaining about the overtime policy. The management finds this astonishing, because their overtime policies are very generous, better than the competition. Still, the linear thinking manager will believe that the overtime policy is the issue. The linear thinking manager might form groups to study it, change the policy, or maybe set up sessions to teach people how good the policy really is. If the manager changes the overtime policy, another set of complaints crop up, this time about the lack of decaffeinated coffee at the water cooler. The systems thinking manager knows that there is a difference between a symptom (the complaining) and the real problem (still unknown). The systems thinking manager will work to understand the problem before addressing any of the symptoms. Usually, if the problem is solved, the symptoms go away.

Figure 5
Linear Thinkers and Systems Thinkers

Linear Thinkers	Systems Thinkers
Break things up to component pieces	Are concerned with the whole
Are concerned with content	Are concerned with process
Try to fix symptoms	Are concerned with the underlying processes
Are concerned with blame	Try to identify patterns
Try to control chaos to create order	Try to find patterns amidst the chaos
Care only about the content of communication	Care about content, but are more attentive to interactions and patterns of communication
Believe organizations are predictable and orderly	Believe organizations are unpredictable in a chaotic environment

Chapter 13

Is it a Symptom or a Problem?

W e can see symptoms. Symptoms are the outward manifestation of a problem. So, when a person has a heart attack, they frequently have pain down their arm. The heart is not in the arm, but that's where the pain is. Dizziness, sweating, nausea, and shortness of breath can be symptoms of a heart attack, but alleviating those symptoms will not help the problem. In other words, when a person has a heart attack, it is not their heart that hurts; other symptoms can alert us to the underlying problem. And we had better not ignore those symptoms! Symptoms are not the problem, but they give us important information about the problem.

The same goes true for organizations. The underlying problem may result in a variety of symptoms that seem distant from the original problem. How do you know if the thing you're seeing is the real, underlying problem, or simply a symptom of an underlying problem? There are no sure-fire ways to tell. There is not a test or a set of questions to determine with 100% accuracy whether the issue is a symptom or a problem. We need to look at clues. There are nine clues below. In your organization, the more true these clues, the more likely it is that you are looking at a symptom and not the underlying problem.

1. *The size of the problem isn't commensurate with the discussion around it.*

   Is the problem too small in comparison to the time and energy it's taking? If people are spending all their time, for example, complaining about the color of the carpet or the shape of their office, that gives you a clue that the complaining is a symptom of another problem.

2. *People don't solve a solvable problem.*

   Is it within the power of the people to solve the problem? But they don't solve it? For example, people complain that there is no decaffeinated coffee at the water cooler. This is a simple problem to solve. But if no one solves it, and then complains that no one solves the problem, this would be a clue that the problem is something else.

3. *The problem won't go away.*
   What has been the history of the problem? Is it a problem that won't go away? Have you tried to solve it and have been unsuccessful? Have you tried to kill it, and like a bad monster movie, it keeps coming back? Does the problem morph into a related problem once you "solve" the original problem? Generally speaking, if you "solve" it, and it comes back, then it's not the underlying problem.

> *Great book on organizational change from a complexity perspective: Edwin Olson and Glenda Eoyang,* **Facilitating Organizational Change.**

4. *Emotional barriers are involved.*
   In the middle ages, prior to Christopher Columbus, people were afraid to sail south of the equator. Contrary to popular belief, they were not afraid of falling off the edge of the earth. (About 500 B.C., the ancient Greeks had proven that the world was round.) The sailors, however, were unwilling to do it. It had never been done before. It wasn't that they all sat around and said, "We're afraid to go south of the equator." The problem was that *they never even thought of it.* It was a stunted imagination. This was an emotional barrier. The same kind of emotional barriers are present in today's organizations. What are the things that are never spoken, never talked about? What things could be done, that if spoken, people would laugh and sarcastically say, "yeah, right." What are the emotional barriers? Where is imagination stunted?

5. *The problem has a pattern.*
   Does the problem have an annual cycle? If so, it may be a symptom. Is the problem predictable? If so, that is a clue that there may be an underlying problem.

6. *The organization has kept the problem around, as a pet.*
   Most organizations need problems. The most common reason that

organizations keep problems is that it gives them something to fuss about. Then, they can fuss about this problem rather than solving the real problem. In a healthy organization, if there is a problem, they solve it. Unhealthy organizations keep problems around as a pet. So, one of the questions to ask is, "Who's benefiting from the problem?" Keep in mind—no one *tries* to keep the problem, and everyone says they want to solve it. No group of people sits in a smoke-filled room and conspires how to cause and keep organizational problems. However, systemic problems emerge, and—beneath the level of awareness—we keep the ones we like!

7. *Other stresses and anxieties are present in the organization.*
   In organizations that are stressed, people find outlets. In organizations that are domineering, for example, people can feel victimized. Most people, when victimized, will complain about other things rather than feeling victimized. Are there other stresses present in the situation? Problems never happen in a vacuum, they always have a context. What is the context of the problem? The more stresses that are present, the more likely there will be a variety of symptoms.

8. *Anxiety is present in the organization.*
   The more anxiety in the organization, the more likely that real problems are hidden, manifested only in symptoms. Staying attentive to the emotional climate of the organization is a helpful clue. Are triangles present? Organizational secrets? Identified patient? These are ways that people dilute anxiety. Are the forces of togetherness strong? Is differentiation low? The more techniques the organization uses to dilute anxiety, the more likely that symptoms, not real problems are manifest.

9. *Linear approaches to management and leadership have been used in the organization.*
   The more linear the approaches to problems, the more likely that there are many symptoms for many problems. A linear approach to problem-solving will move symptoms around. Consequently, an organization with a history of reactive, quick-fix, cause-effect management, is likely to have symptoms moving all over the place.

Remember, making the symptom go away will not solve the problem. The problem will, in fact, be made worse, and erupt in another symptom. But since most linear thinkers don't connect the two symptoms, you may be rewarded for "solving the problem." In actuality, you've simply moved the problem to another place. You may even get promoted for your "fine work." Yet, all you did is make a symptom go away, make the problem worse, and the symptoms erupted somewhere else.

Chapter 14
Principles of Thinking in Complex Systems

1. Linear thinkers are concerned with breaking things up into their component pieces. Systems thinkers are concerned with how the whole behaves.

We all operate with certain philosophies. We are aware of some of our philosophies; others are tacit (one of the hallmarks of maturity is an active reflection on our tacit philosophies, that we may become aware of them). Linear thinkers tend to operate with philosophies of reductionism and determinism. In reductionism, thinkers will believe that the most important step in a problem is breaking it down into its component parts.

Principia Cybernetica Web is a giant encyclopedia of complexity concepts. http:// pespmc1.vub.ac.be/ Default.html

Determinism is to believe that when you have the same input, the same output will occur; everything is predictable and finite. So, if we do a thing the same way once, and get a particular result, if we do it the same way again, we will get the same result. Linear managers love to have equity across organizations, and policies that will cover everyone in the organization, and have procedures for every single issue that could ever come up.

The way organizations work, according to traditional, linear theory, is that if you create certain inputs, then the inputs go into a black box. (The black box is never explained. There is almost a magical quality about the black box. "Something happens in the black box, and we're not sure what, but it doesn't matter either.") Then, out the other side of the black box, comes the result!

Systems thinkers understand that identical inputs do not

always create the same outputs. The steps to create quality in one place will not necessarily create quality in another place. The steps needed to create a product in one place may not be the steps you'll need in another place.

2. *Linear thinkers are concerned with content. Systems thinkers are concerned with process.*

In the previous example of the conflicted family, the linear-thinking therapist will listen closely to the content of the complaints—why Johnny is mouthy, what skills Johnny needs, etc. The *content* was the expressed ideas. A systemic therapist would focus on the *process*—what are the lines of communication in the family? Who has the power in the family? What are the unspoken dynamics going on?

> *Systems is a new way of thinking, especially if you're used to linear thinking. It takes practice and experience.*

In the green and grassy lawn, the content—or the outcome—is the green grass. If you are only concerned with outcomes, you'll want to put on heavy chemical fertilizers with weed killers. If you're concerned about process, you'll want to know how to create a healthy soil. Similarly, if you want to kill mosquitoes and you only care about outcomes (content), then you can pour on the DDT. But if you care about processes and ecosystems, then you have a much bigger picture to look at.

3. *Linear thinkers try to fix symptoms. Systems thinkers are concerned with the processes that underlie the problems.*

Another difference between systems and linear thinkers is the difference between attacking the symptom of the problem or the problem itself. Symptoms are easy to see; real problems are usually disguised. Symptoms are easy to fix; real problems are not. Addressing the process underlying the problem—and not the content of the problem—is counterintuitive (and will get you strange looks from everyone around you).

When organizations are living with a lot of anxiety, they tend to complain a lot. Their well-meaning but linear-thinking leaders may try to "fix" the expressed problem. So, for example, the organizations tend to manifest their anxiety by blaming one of the departments. They say, "if only that department wouldn't get all the resources, the rest of us would have enough." So, the linear thinking leader may pull together vast budget data to prove to everyone that this particular department didn't get more money than anyone else. Will that fix the problem? It may *look* like that fixed the problem, because that complaint will go away. But, another complaint will rise, e.g., "how come we don't have a casual dress day on Fridays?" Same anxiety, different topic. The linear thinking leaders may then institute a casual dress Friday program. The leaders solved another symptom, but the problem remains. Now, the systemic anxiety will simply find a new thing to fuss about—perhaps they'll start complaining about the taste of the coffee at the water cooler. So, linear leaders will then solve that symptom. That symptom goes away, and a new one emerges, on *ad nauseum*.

4. *Linear thinkers are concerned with blame. Systems thinkers try to identify patterns and mutual influences.*

When there is a problem, linear thinkers look to blame something or someone (and usually that blame will have nothing to do with themselves). Typically, they'll blame someone, and they'll attribute the problem to the person's stupidity or they'll use some other pejorative attribution.

When a leader makes an unpopular decision, the followers usually attribute the decision to some aspect of the leader's personality or heritage (e.g., who the leader's mother slept with, etc.). Rarely do followers stop to find out the whole issue. And rarely do the leaders bother to share the whole issue. This principle is not talking about responsibility. It is normal and healthy to hold people accountable for their

> *Blame is the most common manifestation of mechanization. How many times have you heard "This organization would be great if only so-and-so wasn't incompetent"?*

actions. This linear kind of blame is different. It's a narrow perception, a tacit or deliberate lack of information. When there is little information, you can invent a reason why the leader did it. You can blame the leader, and you don't have to feel guilty about it. When you have all the information, it usually makes you much more sympathetic with the predicament of the leader—whether or not you agree with the decision. This is a much more complex set of emotions.

Systems thinkers try to understand the broad picture of patterns, interactions, influences and forces at work in a situation. True systemic thinking is difficult to find. It's a whole lot easier to focus the blame on one person, place, or thing, rather than to look at all the issues... and to look at, God forbid, how my own actions or position in the system might be contributing to the problem.

5. *Linear thinkers try to control chaos to create order. Systems thinkers try to find patterns in amidst the chaos.*

Linear thinkers spend a lot of time trying to control things. They like to control people, events, and the future. The great illusion of course, is that they never had control... they just had the illusion of control. No one can control others. No one can control events. And to the disappointment of many, no one can control the future (perhaps someone should tell that to all those Committees for Strategic Planning).

A manager *can* control how many people go to the break room at one time. But that's about it. Managers can never get people to be committed. Managers can never motivate people who don't want to be motivated. Managers can never force people to "grasp the vision" of the organization. Managers cannot coerce people to be less toxic. Managers can't control the important things.

Systems thinkers look at the forces of societal attitudes, of government, of the environment, of the economy, of demographics, and realize that they cannot control it. They look at the variety of forces in the organization—the personalities, the homeostatic influences, the external forces pushing on it, and they realize that they cannot control it. They see their market—the people they serve, their customers, their clients—and they realize that they cannot control them. All these forces pushing and pulling are rolled together in a giant chaotic stew. All the forces together are chaos.

But chaos always has patterns. Amidst the turbulence of all chaos, are currents of quiet, orderly patterns. One of the keys in leadership is to find the order in all the chaos. We shouldn't try to manage the chaos, because that would be impossible. But looking to find the order is possible, and provides a payoff.

6. *Linear thinkers care about things. Systems thinkers care about interactions and patterns.*

Systems thinkers remember the lessons of subatomic physics: there are no things—only interactions. It is the patterns of relating and behaving that are keys to understanding organization. Events, job descriptions, job titles, and tasks have less to do with how things work than with the patterns and interactions. If the interactions are full of health, wholeness, and integrity, then you will probably have good things going on in your organization. If the interactions are secret, full of anxiety and self-protection, then that will also say volumes about the organization.

> *"Men stumble over the truth from time to time, but most pick themselves up and hurry off as if nothing happened." – Winston Churchill, 1874-1965*

Free Hand-Washing Tips

One of the greatest names in the history of medicine was Ignaz Semmelweis (1818-1865). He was a Hungarian doctor who practiced before people knew stuff about bacteria, infections, and antiseptic conditions. He practiced in a hospital where thirteen percent of the women died shortly after they gave birth. They died of what they called puerperal sepsis, or childbed fever. Today, we know they died from the unsanitary conditions and the passing of bacteria.

In 1847, Ignaz and his friend, Jakob, were conducting an autopsy, discussing the terrible number of childbirth deaths. Distracted by the conversation, Jakob accidentally cut his own finger with the scalpel. He died of an infection a few days later.

Ignaz realized that his friend had the same symptoms as childbed fever. He did some study, did some experiments, and decided that it was the doctors who were transmitting the problem. The medical establishment thought his ideas were lunacy (Joseph Lister didn't start killing bacteria for another eighteen years).

Ignaz found that simply by washing hands and changing bed linen, the incidence of childbed fever could be dropped from thirteen percent to one percent.

Now, you'd think that kind of result would instantly be recognized and immediately put into practice. But in reality, it was decades before hand-washing became commonplace. In fact, he was put into a mental hospital for his strange views (and he was also an ornery cuss and no one liked him). In one of the great ironies of history, he cut his finger while he was in the asylum, and he died of an infection a few days later.

Most of the nineteenth century medical establishment had simply accepted the sad fact that thirteen percent of women will die in childbirth. Ignaz didn't accept that. He looked at the problem in a new way. He asked new questions and came up with new answers.

There are commonly accepted ways that organizations look at their particular problems. There are approved ways of understanding problems. When you look at organizational problems in a new (and unapproved) way, people will think you've gone nuts. People will want to throw you into an asylum until you learn to think right.

Chapter 15
Ten Enemies of Systems Thinking

T he ten statements below are usually enemies of systems thinking. Hearing these statements is not always a sure-fire guarantee that linear thinking is coming. But if you hear the statements below, warning bells should go off in your head, that this may be a linear approach to a complex problem.

1. *"We've got to fix it quick!"*
    This is the proverbial "quick fix" mentality. We see a problem and we react to fix it before we really understand it. There is nothing wrong with quick, assertive action, and a systems response to a problem is not necessarily slow. But doing the fix before the problem is understood is a linear solution waiting to happen.

2. *"Oh, let's just put a band-aid on it."*
    The band-aid solution. The danger here is to make a half-hearted attempt to fix a problem. The danger is to cover up the worst of the symptoms and let someone else deal with the problem.

3. *"We must make the budget by the end of the fiscal year!"*
    Budgets are notoriously linear. The danger is to make decisions based on the money we have, rather than make decisions based on if the idea is a good one. Particularly, making a decision to fix something so that we are "in the black" by an arbitrary deadline is dangerous. While being in the black is a good thing, the last-minute Herculean efforts to be profitable at all costs is near death for systemic thinking. Short-term budget quick-fixes almost always harm long-term sustainability.

4. *"We need to respond immediately!"*
    Knee-jerk reactions and panic attacks create linear solutions. Rather than a calm, reasoned strategy, this panicked reaction is borne out of anxiety and learned helplessness. A non-anxious

approach is the preferred systems response. This does not mean acting slowly. It means approaching the problem with calm rather than panic.

5. *"Who cares?"*
   A lack of curiosity is the death-knell to systems thinking. Curiosity, play, imagination, and adventure are the antidotes to stuck organizations. An apathetic approach, or a plain lack of curiosity is a red flag to a systemic problem.

6. *"We need more information."*
   There is nothing wrong with more information, unless we believe it will solve the problem for us. If we believe that more information will solve a problem, or embolden us to solve a problem, we create an

   > *Enemies of linear thinking: spontaneity, humor, imagination, innovation, curiosity, experimentation, & play*

   environment for linear thinking. More information is a good thing when we know information's place. We (not information) have to differentiate, and we (not information) must have the courage to act.

7. *"Oh, you're just thinking too deeply."*
   Shallow and superficial thinking is everywhere. We learn how to do it by watching the nightly news: All the complex problems of the world are boiled down to a few sound bytes. The accusation of "thinking deeply" usually means "stop thinking differently than me." The reality is that systems thinking is deep thinking, and not everyone likes thinking deeply.

8. *"To hell with the rest of the organization, we must get our own needs met."*
   This is a fortress mentality. In the middle of our organizations, we live in bunkers, protecting our own needs and the resources of our unit. Consequently, we end up thinking of win-lose strategies, and strategize about how to get more for ourselves. Again, a red flag for linear thinking.

9. *"We can't have any conflict about this."*
   This is a peace-at-all-costs mentality. People in this mindset will do anything to keep the peace. Edwin Friedman calls this "peace-mongering," and it is usually a function of poor differentiation

and high anxiety. Peace-mongers will avoid conflict, suppress conflict, and mask conflict. Consequently, real issues do not get discussed.

You've heard that "Knowledge is power." In systems, we would rephrase that to say, "Knowledge is sometimes used to dominate people."

10. *"You will do it this way and you will enjoy it!"*
Domineering interventions by authoritarian managers are big red flags that linear thinking is coming. Wisdom is collaborative, and prima donna managers who force their will on the unsuspecting populace are typically dispensing linear thinking.

Figure 6
Process Adaptive Systems in a Nutshell

If you ask this question	Then you can learn
What is the level of anxiety?	How ready the system is for change. How toxic is the environment. The predilection for triangles, poor boundaries, and symptom-solving. How sticky the changes will be.
What are the patterns of interaction?	When you can ignore the content of the issue, and just deal with process. How people will respond to you or your ideas. What are the points of greatest leverage?
What is the level of differentiation?	How conflicted the organization is. How many toxic forces are present in the organization. How strong the forces of togetherness are. How well the organization deals with stress. How clear the boundaries are.
How linearly do the leaders think?	How snarled the organization is in its own symptoms. The ability of the organization to adapt to its environment. The frequency with which problems have been blamed rather than solved.
How preoccupied is the organization with certainty and control?	How adaptive the organization can be. How creative the organization is willing to be.
What are the tools the organization uses to maintain homeostasis?	How difficult (or easy) change will be. How long it has been since the organization has changed. The level of sustainability and adaptability.

Chapter 16

Making the Jump to Systems Thinking

When Albert Einstein began to play with quantum physics, he didn't like it. He spent a few years trying to disprove quantum physics, because "it didn't make sense." But in the end, Newtonian physics couldn't answer his questions. His only choice was to become a quantum thinker. This didn't mean that Newtonian physics went away. It simply meant that there were many occasions when he had to use quantum, rather than Newtonian physics.

In order to become a systems thinker, you need an unending curiosity, long-term persistence, and a dissatisfaction with linear thinking.

People don't become systems thinkers because systems thinking is so cool. People move to systems thinking because they are dissatisfied with linear thinking. People move to systems when they discover that linear thinking won't answer their questions.

However, many people never discover that linear thinking doesn't work. They tend to be people who don't think deeply enough or reflect long enough. They are the people who have made up their mind, and they don't want the facts to confuse them. They see phenomenon, and then force the interpretation into their linear paradigms. They are the people who enjoy the comfort of opinion without enduring the discomfort of thought.

One of the problems that linear thinkers have is the time delay. If enough time passes between the action and the effect, they forget that there was a nonlinear cause.

Moving to systems takes persistence. Most people think linearly; so, being the sole systems thinker in a linear thinking organization can be a lonely place. People will not understand you. You'll feel like you're walking

Figure 7

On the Road to Becoming a Systems Thinker

Instead of this	Try this
Instead of blaming someone, ask...	"What are the influences on that person?"
Instead of saying "I know the answer," say...	"I have another perspective on the issue."
Instead of thinking you know the answer...	Always be curious; always be looking for evidence to confirm your theory and evidence to disconfirm your theory.
Instead of focusing on one item...	Look at all the variables that affect that one item.
Instead of looking at the content of what people say...	Look for the process of what they say. How are they saying it? What are they not saying? What are the common themes in the content?
Instead of looking at fussing, complaining, and gossip...	Look at what is motivating the fussing, or if the fussing masks a deeper problem.
Instead of just looking at what individuals are doing...	Also look at the dynamics of the system; what forces are pushing individuals toward one thing or another?

in a giant dog pound... whenever you talk, people will tilt their head and stare at you, confused.

Moving to systems thinking also takes an unending curiosity. Most people don't realize that linear thinking isn't working simply because they aren't attentive enough. They don't listen close enough; they aren't curious enough. Curious people are always trying to figure out the dynamics of the situation, learning as many of the variables as possible. The linear thinker would say, "Curiosity killed the cat." The systems thinker would say, "Curiosity killed the cat, but satisfaction brought him back." Shades of Schrödinger.

Of course, many people have thought systemically all their lives. Many of those people don't even know that what they are doing is labeled "systems." But for the vast majority of us, the jump to systems thinking requires time, effort, practice, curiosity, and intentionality.

Free Martial Arts Tips

If someone approaches you with evil intent, the most common reaction is to try to meet that force with equal or greater force. The linear approach to fighting is to go head-to-head, toe-to-toe, until the other person backs down. Whether we are talking about individuals or nations, the typical way to engage in conflict is to meet the opponent head-on.

Many martial arts practice endless punches and kicks, with the hope of being stronger and better prepared than the opponent. In classic Karate, for example, if someone is rushing at you, you execute a defensive side kick. The side kick is designed to stop an attacker in his tracks—knocking them back with an explosive force.

The Japanese martial art of Aikido, on the other hand, does not engage in nose-to-nose conflicts. The Aikido stylist uses the attacker's own weight and momentum against him. For example, if someone is rushing at you, you step to the side, and touch him on the back of the neck. The extra force will cause the attacker to go faster than he expected, and he will lose his balance and plunge face-first into the ground.

When a person attacks, the Aikido stylist may move toward the attacker, not away. The Aikido stylist may grab the attacker's chin and gracefully pull the attacker in a slightly different direction. By adjusting the direction of the attacker, the Aikido stylist suddenly has complete control of the attacker. By keeping the attacker close, they can manipulate the direction in which the attacker is moving.

Many experienced martial artists have difficulty learning Aikido. In Aikido, you learn to use less force, not more. You learn to control the attacker's movements rather than subdue him. You learn to move toward the attacker, not away from him. You learn to move in circles rather than straight lines.

When you start learning how to operate in complex systems, you have to re-learn everything you know. You have to unlearn things that seem like common sense. You have to watch patterns that you never knew existed. You have to do things that seem counter-intuitive.

And, whether we're talking about Aikido or complex systems, people will not understand what you are doing. But it works. And it looks cool.

Chapter 17

Summary

There is nothing wrong with linear thinking. It is a level of what's going on—but it's only *one* level. Those enamored with linear thinking will miss out on most of the reality around them.

We are used to thinking in traits—that person is incompetent, that person is addle-minded, that person is obtuse. This "trait approach" to life might be called *psychological thinking*. We need to get away from seeing interactions and dynamics through "trait" glasses. We need to decrease our psychological thinking and learn to think more biologically. In biology, fish don't have traits, deer are not incompetent, and wolves just do what wolves do. But, if you affect one part of a biological system, then you affect all parts of the system.

Organizations are not static and unchanging. The environment where organizations sit is chaotic and dynamic. Organizations are not machines—they are ecosystems. The best way to survive and thrive in a changing world is to think biologically!

Free Snake-Handling Tips

Steve Irwin, the famous crocodile hunter, was recently in the Australian Outback looking for snakes. He found, of course, one of the most venomous snakes in the world. And, of course, he picked it up.

He had the snake by the tail, and the snake began to thrash wildly trying to bite him. The snake was so poisonous that if it would have succeeded in biting him, Steve would have been dead before he hit the ground. He allowed the snake to thrash for a while, but then he put his hands under the snake, and allowed the snake to move through his hands. He moved his hands to keep the snake in place—in a sense, stroking the stomach of the reptile. Soon, the snake was calm and was no longer trying to bite him.

The snake, by some quirk of nature, has the brain of a reptile. It can't decide "should I bite Steve or not?" The snake reacts instinctually. It tries to bite. It can't think about not biting. Operating only on impulse and instinct, it just bites. It can't think.

Poorly differentiated people operate on instinct. They bite, they attack, they have anxiety tantrums, they freak out. An instinctual person can't think about not attacking, they simply attack. It's not about the content of the issue, it's about instinct.

Steve Irwin was calm with the snake. He was in touch with his own anxiety about the snake. But he calmed the snake by staying in touch with it. Had he tried to out-thrash the snake, he'd be dead.

When we work with instinctual people, we usually try to meet their attack with equal or greater power—in good linear thinking fashion. Steve Irwin would have been justified in being full of anxiety about handling that snake. He was handling a deadly snake miles from a medical facility—if ever there was a justifiable time to be nervous, it was then. However, his calm was a prerequisite to calming the snake.

If the poorly differentiated person is dysfunctioning all over the place, we can meet their anxiety with calm. We can stay in touch with the person, and our non-anxiousness can calm them. In fact, calming ourselves is a prerequisite to calming another. If we try to meet their instinctual response with our instinctual response, things will escalate until someone sinks their fangs into someone else.

Just for the record, Steve knows how to pick up poisonous snakes. Don't try this at home.

Part 3: Ant Hills and Organizations: Introduction to Complex Adaptive Systems

Chapter 18

Ants

No ant knows how to make an anthill. Deborah Gordon, in her book *Ants at Work* (1999), chronicles her 15-year study of ants. In the wilds of the Arizona desert, she has excavated anthills, used fiber optics to see ant chambers, marked ants with paint, observed them, studied them, and bothered them. Her mission has always been to "find a pattern and then figure out how and when it changes" (p. ix).

No ant directs the colony. Contrary to popular belief, the queen is not in charge. She simply sits in the bottom of the anthill and lays eggs. She directs no one. There is no management in the anthill. There is no ant leader. There are no old, wise ants that mentor the young, inexperienced ants. Complex anthills simply emerge.

There are almost 9,000 species of ants in the world.

The hills of the red harvester ant measure about a meter across. Underneath the soil, there is a cone-shaped colony of tunnels and rooms, going down about two or three meters. In a mature colony (five years or older), about 10,000 ants make their home there. An anthill survives about 15-20 years (the lifespan of the queen). A normal ant lives about one year in the laboratory, and usually only a few weeks in the desert. When the queen dies, and no replacement ants arrive, the anthill dies off.

The colony begins when a winged queen sets off from another colony. A number of male ants mate with her (kind of an ant orgy). She ultimately shakes them off and tries to find a place to start her own colony. She loses her wings, and then digs a hole in the ground. Ants are hatched, start digging, and then the queen lays more eggs, and those newborn ants continue refining the anthill. Of course, only a very small percentage of queens succeed in starting their own colony. Most are eaten by birds, lizards, or other ants.

Ants have tasks, and they can keep that task for life. The first kind of ant task is the patroller, who goes out early in the morning. The patrollers survey the landscape for danger and food, and they will inspect other ants to determine who is a part of the nest and who is a "visitor."

Foraging is the second ant task. Foragers come out early in the morning and pick up the food that the patrollers have identified. The foragers will go only to the areas that the patrollers have identified (patrollers apparently mark the designated areas with a chemical).

The South American Army Ant does not have a nest; they travel around and eat anything they can get their mandibles on.

The third task is nest maintenance. These workers dig more tunnels, tend to the queen and her brood, and make sure the insides of the nest are functioning smoothly. Nest maintenance workers rarely go outside the colony. The fourth type of ant task is the midden work. These ants carry ant dung and dead bodies out of the nest (this is the job you don't want).

The foragers bring the food back to the anthill and drop it into the nest opening. The nest maintenance workers move the food to one of the rooms in the colony, and stack it neatly for when it is needed. The foragers occasionally have to fight neighboring ants for food, but usually each of the anthills has their own territory. There seems to be a small percentage of the foragers who do the fighting (karate ants).

About twenty-five percent of the colony are foragers and patrollers (those who work outside the nest). About seventy-five percent work inside the nest: midden workers, nest maintenance workers, and the reservists who work when needed.

Ants are not born to be foragers, patrollers, or midden workers. No one assigns them their task. Under normal circumstances, ants don't change their task. But when special circumstances dictate, they can change their tasks.

When there is an overabundance of food, many of the nest maintenance workers will switch to foraging. When there is a sudden need for nest maintenance, the reservists appear to help with the work (for some reason, foragers and patrollers don't return to nest maintenance).

So, if the patrollers discover a large quantity of food, the foragers will retrieve it (interestingly, if high-quality food is placed near the anthill after the patrollers have finished their duty, the foragers will walk over the good food to get the low-quality food that the patrollers have identified). The more food that is brought in, the more foragers appear to help. Nest maintenance workers will switch tasks to help with the foraging. No one tells them to do this.

So how do the nest maintenance workers know how to switch to foraging?

Apparently, the ants rely on whom they meet. If an ant meets a forager, and then another, and then another, the ant decides that there must be a lot of food out there, and so the ant will turn to foraging.

On the other hand, if the foragers come back empty-handed, less foragers appear. In this way, the anthill can conserve its resources. So, when the foragers come back empty-handed, this is a clue to the other ants to lay low. No ant tells another ant to stop foraging. There are no old, wise ants telling the young foragers to lay low. Ants do not talk to each other in an interaction.

In case you're wondering, an anthill is an example of a complex adaptive system.

Ants know what to do by the *pattern of* interaction. It is the repeated bumping into an ant with food that will motivate another ant to go a-foraging. *It is the pattern of interaction, not the interaction itself that communicates the need.*

Gordon has done a wide variety of experiments. She put a pile of toothpicks by the opening of an anthill to see the colony's reaction. A large number of nest maintenance workers, and a bunch of reservists who had been drafted, came out and moved the toothpicks.

Gordon did a similar experiment on the nest maintenance workers. She put a steel pipe in the sand near the anthill. She tied a steel wire to the top of the pipe, and on the other end of the wire hung a cardboard tube so that it dangled precariously over the anthill opening (ants find this annoying). Within minutes, the nest maintenance workers scaled up the pipe and tried to cut the wire with their mandibles. Unable to cut the steel wire, they ultimately gave up, and ignored the offensive structure.

You cannot control an anthill. You can annoy an anthill, but they will adapt. Ant spontaneously adapt to new situations. Ants are highly organized, but without leadership, direction, or management. Ants organize themselves without any hierarchy or central controlling authority. An anthill is a *self-organizing system*.

You cannot find out much about an anthill from looking at an individual ant. Ant behavior makes sense only in the context of the colony. Only the pattern of behavior makes any sense. If you only looked at one ant, the ant behavior would appear random. When you see the whole colony, the behavior creates a pattern and has a purpose. This purpose emerges from the whole colony, not from any individual. Ant behavior is *emergent*.

Patroller ants form zigzag patterns in their attempts to locate food and watch out for intruder ants. They wander around, and come back to the nest empty-handed. When viewed from their singular perspective, their behavior appears random and meaningless. But without the patrollers, the foragers would not pick up food, and the anthill would die. So, when viewed from the perspective of the whole system, the apparent chaos has a pattern. In nature, *chaos has a pattern*.

An anthill is not just a single system; it is an interconnected group of systems. What happens to the foragers affects the nest maintenance workers, which affects the midden workers. The patrollers are affected by the ants of the neighboring anthill, and the colony is affected by predators such as ant-eaters. An anthill is an interconnected group of systems, called a *complex system*.

Ants adapt to their circumstances. If they cannot cut down annoying cardboard tubes, they ignore it. If they realize a fungal infection in their anthill, they will leave and create a new anthill. If their foraging area conflicts with another anthill's area, both anthills will create new areas. Anthills adapt to challenges, threats, and circumstances. An anthill is an *adaptive system*.

Chapter 19

Why Complex Adaptive Systems?

L ife seems to be getting faster and more chaotic. In our organizations, there is more and more to do. Tighter budgets. An increasing plethora of cultures. Technology breaking down barriers. Economic upswings and downturns. In an environment of unlimited possibilities, it seems that there are no boundaries. In faster, more unpredictable times, many leaders agonize about how to get control of their organizations.

It is as if our organizations are boats in a white water rapids (Vaill, 1991). Our little organization is tossed to and fro by the wild currents and splashing water. Our little boat is going faster than we can control—and there is no calm water in sight. We try to strategically plan, set goals and objectives, create scenarios to understand the future—all in a futile attempt to control the boat in the white water rapids. But we can never control the white water rapids. The best we can do is steer our boats in a way that gives us a small ability to navigate away from the dangerous rocks.

One of the classic books on complex adaptive systems is by the Nobel-prize winning physicist, Murray Gell-Mann. The book is called **The Quark and the Jaguar.**

Complete control is a myth.

It is impossible to control an organization in times that are increasingly chaotic. You can't control the white water rapids—the best you can do is sail through it effectively. You sail through it effectively by understanding the flow of the water, navigating through the currents and eddies (and avoiding the large rocks).

Scientists have been studying the phenomenon of *chaos* and *complexity* for several years (Gleick, 1987). It seems that in nature, things that appear chaotic—like white water rapids—always have an underlying order.

Perhaps, then, understanding the properties of chaotic systems and their underlying order can be helpful in sorting out the perceived chaos in our organizations. Perhaps, if we look carefully, we can see some patterns that will help us lead our organizations more effectively.

This is the study of complexity.

Biological species, like organizations, emerge, grow, maintain themselves, and sometimes become extinct. Species, like organizations, work to survive and thrive. Sometimes external things happen to the species that harms their long-term viability. Sometimes the species do things to themselves that hinder their long-term viability.

The word complex *comes from the Latin word* complexus, *meaning, "to weave together."*

Species that last a long time have learned to adapt to their environment as it changes. They change themselves to meet the new challenges of the new environment.

Perhaps we can learn about organizational growth, death, and viability by studying biological systems. Perhaps, if we look carefully, we can learn something about biological systems that will help us lead our organizations more effectively.

We don't study complex systems because we occasionally find some analogies to organizational life. Both a biological ecosystem and a human organization are complex systems. There are similarities in how the two types of systems function. The common principles of complex systems—be they ants or people—can be instructive in our understanding of how we function.

This is the study of complex adaptive systems.

Chapter 20

Characteristics of Complex Adaptive Systems

Complex and Adaptive

A complex adaptive system is an interconnected group of systems that organize and change without the direction of a central authority figure. They organize and change to be a better fit with the environment.

A complex adaptive system consists of a number of components, or agents, that interact with each other according to sets of rules that require them to examine and respond to each other's behavior in order to improve their behavior and thus the behavior of the system they comprise (Stacey, 1996).

A complex adaptive system is composed of interacting agents following rules, exchanging influence with their local and global environments, and altering the very environment they are responding to by virtue of their simple actions (Santa Fe Institute, 2000).

Note the difference between complicated and complex.

Complex suggests that it is an interconnected group of systems. This is different than a *complicated* system, which suggests lots of activity, with no real interconnected purpose. *Complex* suggests that there are nonlinear (not cause-effect) relationships that occur across the system.

Adaptive suggests that the system changes over time, that it adapts to its surroundings so that it can survive and thrive. The process of adapting, of course, changes the environment, which facilitates more adapting in the system.

An anthill, for example, is a complex adaptive system.

In a complex adaptive system, it is impossible to count the number of agents, variables, or parts of the system. This phenomenon is called the *indefinite number of agents*. In many cases, the agents are too numerous too mention. In most systems, there are simply too many forces and influences acting on any part of the system too count.

Another way to look at the indefinite number of agents is that, in a complex adaptive system, the boundaries are often blurry. It is sometimes impossible to tell what is a part of the system and what is not a part of the system. The boundaries of most ecological systems are blurry. For example, a lake in Northern Minnesota: where are its boundaries? The shore? Where exactly does the shore start and end? Do you count the cattails on the edge of the shore? Are the frogs that live in the mud a part of the lake? How about the mist that rises from the lake in the morning?

- A lake is a complex system. There are many influences on the lake—temperature, incoming water, wildlife, pollution, humidity, human use, invasive species, and so on.
- A jungle is a complex adaptive system. There is a wonderful amount of biodiversity in the tropical areas around the world.
- The weather is a complex adaptive system. There are a tremendous number of agents that affect and influence the weather.
- A traffic jam is a complex adaptive system.
- Your organization is a complex adaptive system.
- Your immune system adapts to a myriad of invaders and viruses—killing pathogens that it has never seen before.
- The economy is a complex adaptive system. Some people have suggested that understanding more about complex adaptive systems will help you understand the stock market better. Who ever said learning wasn't profitable?
- Families are complex adaptive systems.
- A pot of boiling water is a complex system, but it is not adaptive. It does not "learn" from its experience.
- Our galaxy of 200,000,000,000 stars is a complex system. Many forces act on it constantly, and there is a complex interplay of many forces and processes. Its structure is shaped by forces like gravity and dark matter, but it does not adapt to its surroundings.

It does not adapt to be better fit to survive. Galaxies do not adapt—they evolve. They are born, go through a life cycle, and ultimately die—they evolve over time. A galaxy then, is a complex *evolving* system.

One of the central issue in a complex adaptive system is its ability to learn. To understand a complex adaptive system, we must follow the flow of information into the system, and then see what happens as a result of the information. The system's ability to learn, and therefore adapt, is the critical component of a complex adaptive system.

Like it or not, your organization is a complex adaptive system. You can conduct yourself as if the organization were a machine (this is what most people do), and you'll fail to get all the potential out of the organization. If you treat it as a complex adaptive system, you'll be applying the appropriate kind of pressure for the appropriate issue.

Learning

When a system adapts to make itself more fit, we say the system has *learned*. Learning does not have to be an intentional or a sophisticated process.

* When ants find a new home due to the fungus in their current home, that represents learning. The ant "knows" that the fungus would kill them if they stay.

Some authors that write about learning organizations are Peter Senge, Chris Argyris, Sarita Chawla, Peter Kline, and Peter Vail.

* When a tree bends its branches to reach more sunlight, it "knows" that it needs the sunlight to stay healthy. The tree has learned to seek more sunlight.
* When Minnesota timber wolves (the wolf, not the basketball team) stay in their own territory, and refuse to go into another wolf pack territory, they have learned a trait that has helped the species survive. Too many wolves cascading into each other's territory means lots of fighting.

Learning happens in a variety of different ways. In the case of a biological species or an organization, there is the phenomenon of natural selection. In natural selection, it is not the strongest who survive, or those who have the coolest product line, but those who have the healthiest fit with the environment.

Another way systems learn is when they stay attentive to feedback. Feedback is the information that comes back to the system. Feedback is not only verbal; in fact, most feedback is non-verbal.

- When we fill a glass of water, our eyes tell us the glass is full. This information is feedback.
- When we make a leadership decision, we will discover a variety of resistances going on. This resistance is feedback.
- When we read for a long time, we get a headache. This feedback tells us that we are straining our eyes (or that the subject is boring and obtuse).
- When toddlers learn to walk, they experience clear and unmistakable feedback every time they make a walking mistake (they fall down go boom). Toddlers learn to walk by being attentive to the feedback.

There is always feedback within the system. In some way, shape, or form, the system gets feedback about its behavior. It rare cases, it will be straight-forward and clear, like escalating purchases of a new product. Usually, however, feedback is ambiguous, like my art professor used to say, "well, *that's* interesting." Sometimes feedback is coded, such as in a family, you may hear, "*someone* forgot to fill the ice cube tray."

Feedback can be amplifying, meaning it supports the action. When you have amplifying feedback, the chances of the action happening again are enhanced. When most people eat chocolate, it is such a supernaturally breathtaking experience, that the chances of it happening again are enhanced.

The opposite of amplifying feedback is damping feedback, which pushes toward homeostasis—toward the new action *not* being repeated. If you were to put your hand in a hornet's nest and wiggle it around, you would suffer clear and unambiguous damping feedback. It is likely you wouldn't perform that action again.

Amplifying feedback says, "Nothing ventured, nothing gained." Damping feedback says, "a bird in hand is worth two in the bush." Amplifying feedback make us want to do more. Damping feedback makes us want to do less.

How do organizations learn? They learn when they make changes in their conduct as a result of changes with the environment. Now, we're not talking about knee-jerk response. We're not talking about when someone in tennis shoes slips on the floor, and sues the organization, and now no one is allowed to wear tennis shoes anymore. This is not an adaptive response. That kind of panic-stricken, anxiety-ridden response does *not* make the system better able to function in a changing environment.

An adaptive response is to alter a service or product line as the market needs change. Meeting a need, without a vast change in bureaucratic structure is an adaptive change.

In organizations, there are anti-learning forces. These are the forces that make it more difficult to learn or to adapt as a result of the learning.

- An anti-learning force is in an organization that values internal bureaucracy more than attentiveness to the environment.
- An anti-learning force is in an organization that values tradition more than adaptability.
- An anti-learning force is in an organization that values correct procedure more than experimentation.
- An anti-learning force is in an organization that values equity more than opportunity.
- An anti-learning force is in an organization that relies on the safety of hierarchy rather than trust the judgment of peers.

Fitness

Imagine a forest that had no animals, grass, or plants—just thousands of trees of the same species (never mind that such a place does not exist in nature). This hypothetical forest is not complex, because there is only one species of tree, and nothing else goes on in this forest. There is not a group of interconnected systems, just one simple system. This is a simple adaptive system. Simple systems tend to be fragile and vulnerable.

Imagine that this one-species forest was populated with American Elm trees. If Dutch Elm disease (a nasty fungus) begins in the forest, within a couple years, the entire forest could be destroyed. On the other hand, a forest with a wide number of species and other biologic activity would hinder the Dutch elm fungus. The disease would spread more slowly, and perhaps die out. So, the more diverse the forest, the more resilient the forest. Diversity in a system is usually associated with better health. Diversity means "many different kinds of things" are a part of the system.

The term *diversity*, in our contemporary usage, often refers only to an ethnic diversity. In the complex adaptive system usage, it has a much broader definition. Think biologically.

- Diversity means a diversity of ideas.
- Diversity means to fully accept and embrace new ideas, even the ones you decide not to take (in other words, you don't implement all the new ideas, but you always accept the new ideas).

> *Here, organizational diversity means to respect the diversity of ideas; to understand that organizations need many voices to remain fit; to respect people even when we don't take their ideas.*

- Diversity is to respect new and innovative ideas. Diversity is to create an environment where new ideas are easily conceived and generated.
- Diversity is to never surround yourself with yes-men and suck-ups, but to surround yourself with people who will challenge your ideas and tell you what they really think.

If a system can sustain itself and reproduce, surviving and thriving over many generations, it is said to be a *fit* system. Fitness is a term closely associated with health of the overall social system.

Complexity and Chaos

Chaos, as it is used here, does not mean *confusion* and *anarchy*. It is helpful to think in terms of a spectrum line… on one side is a completely predictable phenomenon. The other side of the spectrum is a phenomenon that is completely unpredictable and random.

Chaos is right in the middle. On first glance, chaos appears random and disorderly. But upon a closer look, there is an underlying order. Chaos has a pattern.

- If you look at the behavior of a single patroller ant, its behavior will seem random. But when you see all the patroller ants together, you see the meaning behind the activity. You see the pattern.

- If you were to see a Minnesota timber wolf raise its leg and urinate (the wolf, not the basketball team), you may think it sprayed a random spot. However, if you were to watch the wolf pack over the course of a season, you would see that the wolves have perfectly marked the boundaries of their territory. When you see the whole, you see the pattern.

- When you look at the stars in the night sky, you see a truckload of stars that seem to be thrown there at random. But if you could stand outside our galaxy—far enough away—you would see a perfect pinwheel pattern. It seems like chaos on first glance, but it has a pattern.

So, *randomness* is a term that means completely unconnected activity. It is completely unpredictable and irreducible. *Rigidity* is a term that means completely predictable.

Figure 8
Chaos

Rigidity	**Chaos**	**Random**
• *Highly structured*	• *Appears random, but on closer examination, a pattern appears*	• *Completely unpredicatable*
• *Completely predictable*		• *Constant change*
• *No change, or very predictable change*	• *Activity is predictable once the pattern is found*	• *No rhyme or reason to activity*
• *Usually managed by a strong external force*		
• *Rigid*		

Chaos appears random at first glance. But upon closer examination, chaos is patterned and orderly. Chaos is the natural state of ninety percent of the things in the universe. Chaos is often unpredictable to the inattentive eye—but it becomes clear when you see the whole. The problem is, of course, that we are rarely privileged to see the whole phenomenon.

When we look at the chaos of our organizations, we see lots of activity. We see personnel changing, the market shifting, and the workload increasing. We see a faster pace of change, a need for a quicker response time, and more information to gather before decisions can be made effectively. We see more needs, more opportunities, but have less resources. Leadership in this chaotic environment does not mean to control it. Leadership means finding the patterns, and guiding the organization through the patterns.

A couple good books on complexity in organizations are: **The Soul at Work: Embracing Complexity Science for Business Success,** *and* **Surfing the Edge of Chaos: The Laws of Nature and the New Laws of Business.**

Complexity is the umbrella term for chaos and nonlinearity. Complexity science is the study of systems that are complex and adaptive.

Nonlinear Relationships

Linear interactions are simple and straightforward. They are cause and effect. One action stems from one cause. One action will create one other action.

Nonlinear relationships are interactions that are multi-causal, multi-effect, and multi-influencing. It's difficult—and sometimes impossible—to follow the nonlinear relationships and their effects when we take an action in a complex system. Nonlinear interactions are unpredictable. In an organization, any action will have many effects—some intended and some unintended. As Peter Senge (1994) says, "you can never do just one thing."

In organizations, we spend a great deal of time in strategic planning. We plan what the future will look like. However, it is impossible to know the future. While we might try to predict a future state, it will be impossible to know the outcome, much less the forces and influences that affect the project along the way. Change agents need to envision multiple futures—without a knowledge of which future will become reality (Eoyang, 1996).

Another example of a nonlinear relationship is called the *butterfly effect*. To explain this term, James Gleick (1987) recounts the story of Edward Lorenz, a meteorologist in the 1960s. Lorenz designed a way to create scenarios for weather, using a new machine called a *computer*. He fed data into the vacuum-tubed machine, and the computer told him what the weather would be like over the next few days. Of course, two sets of identical data would produce the same result. However, when the data was entered that had an extremely miniscule change—the equivalent of a small puff of wind—within a few days the weather result was completely different than the original scenario.

This phenomenon is called the *butterfly effect*. It suggests that if a butterfly in the North America flaps its wings, it will rain in China instead of be sunny. An extremely small effect—the flap of a butterfly wing—will produce nonlinear effects that bounce around the system, creating a multitude of effects. Scientists call this *extreme sensitivity to initial conditions*.

> *Several million people still live in the Chernobyl-contaminated area in Belarus and Ukraine.*

We see the butterfly effect in systems when a small change, a seemingly insignificant innovation over a long period of time, creates a large effect.

- In 1955, Rosa Parks refused to give up her bus seat to a white man. Who would have thought that this would eventually become the civil rights movement?
- In the 1960s, two universities in California worked on a communication program for computers. Who would have thought that this would eventually become the Internet?
- In 1986, a low-level official from the Soviet city of Kiev asked the local nuclear power plant *not* to conduct the proposed safety test due to higher than expected energy use. Who would have thought that this was the first of a series of small events that resulted in the nuclear disaster at Chernobyl?

Self-Organizing Capacity

Perhaps the most fundamental concept of a complex adaptive system is its capacity to self-organize. A perfectly self-organizing system will adapt to the environment without directions from leaders, without instructions from the central authority, and without preplanned blueprints for change.

A self-organizing system organizes to meet various challenges in the environment. This is done without the instruction or direction of a hierarchical leader. It is done without anyone saying, "we should reorganize."

- Anthills self-organize. Ants just kind of know what to do. No ants give orders, but they can create extremely complex structures.
- Environmental ecosystems adapt and self-organize, unless there is too great an incursion (usually by human interference).
- Traffic jams are complex adaptive systems that usually self-organize their way out of existence.

If a forest catches fire, animals will run out of the forest to save themselves. The "running away" is not a self-organized, adaptive process. When a system is forced to do something, it is called direct adaptation or *forced behavior*. In forced behavior, there *is* an external force causing the adapting. In a self-organized adaptive process, on the other hand, there is no external force pushing the change. Forced behavior frequently, but not always, hinders the fitness of a social system.

Forced behavior and *self-organizing capacity* are almost always mutually exclusive. Forced behavior risks the loss of a system's creativity, health, and fitness. Consequently, the manager who likes to give orders and make decisions for people should not wonder why people will not make a move without the say-so of the manager. Forced behavior usually kills self-organizing capacity.

Organizations are likely to become self-organizing when people have a great deal of ownership in their task, when differentiation is high, when anxiety is low, and there is a great deal of trust between the members and the leaders. People are empowered to make decisions regarding the direction of the tasks, and they have the responsibility and accountability for their work. Organizations that are high in self-organizing are playful places, where innovation and creativity are grand adventures; mistakes and failures are not only tolerated, they are expected. When this kind of self-organization occurs, there is a good chance the organization will thrive.

In a rapidly changing world, there is frequently no time for management to make slow and ponderous changes. Organizations must adapt quickly to rapidly changing circumstances. Organizations with over five employees are too complicated for one person to know everything. Consequently, if organizations want to stay in a co-evolving state with the environment, then they must rely on their adaptive, self-organizing capacity.

> *Organizational stuckness is another concept from the work of Edwin Friedman. Systems can get stuck, and then they are unable to adapt. Playfulness and innovation are the antidotes to stuckness.*

Emergence

The three-pound bundle of nerve cells we have inside our craniums is a complex adaptive system. The brain constists of cells called *neurons*, over ten billion of them. A neuron carries a bioelectric charge; that's all the neuron does. If you took the brain, and broke it down into its component parts, all you would find are individual neurons. Studying a single neuron will not tell you very much. You cannot learn the brain's capability by studying the brain's component parts. However, when you put all of these neurons together, a new quality emerges—something that can discuss philosophy, epistemology, and ants. This is the quality of *emergence*.

> *College professors and other animals are more than the sum of their parts. We call this quality* **emergence.**

Emergence is a phenomenon that grows out of the collective behavior of the system. The pattern that emerges cannot be gleaned from an analysis of any one of the parts. Only when the system operates as a whole does the behavior emerge. Emergence, sometimes called *nonsummativity*, is captured in the Aristotle's systemic phrase, "The whole is more than the sum of its parts."

Analysis (breaking things down into their component parts) is a tool, but a limited tool. It gives one layer of understanding. If a learner only performs analysis, the learner will miss the majority of the phenomenon.
- Analysis tries to learn about water by studying hydrogen and oxygen.
- Analysis tries to learn what it's like to walk barefoot through a yard of lush, green grass by looking at a grass seed.
- Analysis tries to learn what an organization is like by reading job descriptions.
- Analysis tries to learn what the system will do by examining the components.

Boundaries

A boundary is a change from one system to another. Boundaries separate things.
- Skin is our boundary that keeps out the world.
- The ozone layer (what's left of it) is the boundary that keeps out the sun's harmful rays.
- The membrane of a cell is the boundary that protects the cell and keeps it intact.
- A good job description is a boundary that helps the organization know how one person's job is different than the next person's job.

Boundaries have two functions. 1) Boundaries keep things intact. If we didn't have skin, for example, we'd be leaving a trail of internal organs wherever we go. Boundaries hold the system together. 2) Boundaries let some stuff in and keep other stuff out. This quality is called *permeability*. The boundary decides what is "a part of us" and what is "*not* a part of us."

Sometimes boundaries must let information or material through. The cells in your body need get nutrients from the blood into the cell and must get waste products out of the cell. This type of boundary is called a permeable boundary. The cell wall acts as a filter, allowing good stuff to get in but it will keep the bad stuff out. If your cell walls were slightly more permeable than they are, water would rush into every cell in your body, making them explode. That would be a bad thing.

Impermeable boundaries, on the other hand, do not let information or stimuli pass through it. These kind of boundaries are useful when you are guarding nuclear secrets or handling toxic waste. Sometimes families temporarily close their boundaries during stressful times; this helps them cope better with the stress. In organizations, impermeable boundaries are important when the external influence would be toxic to the system.

Indistinct boundaries are the opposite of impermeable boundaries. Indistinct boundaries let everything through. It is almost like having no boundary.

In most organizations, impermeable boundaries will create a stagnant system. When no information gets into the organization, there is no learning. With no ability to see the environment or know what it is doing, there can be no adapting. This means that there will quickly be a mismatch between the system and the environment. In biological species and organizational systems, this usually means death.

There are times when rigid boundaries are important, times when permeable boundaries are preferred, and times when indistinct boundaries are needed. Deciding the permeability of the boundaries is a critical task of the leader.

Problems usually occur on the boundaries. That does not mean they start at the boundary. They show up at the boundary.

- Tidal waves are only a few inches high in the middle of the ocean. But at the shoreline, as the water gets shallower, the wave increases. It is at the boundary of the ocean—the shallower water at the shore—where the wave becomes a problem.
- When we get the flu, a virus penetrates the cell wall, injecting its own DNA into the cell. The cell wall did not keep out the viral invader. The boundary failed.
- When the earth's ozone layer thins, and develops a hole, the rays of the sun can create skin cancer and a host of other problems. The problem did not start at the ozone layer; the problem started as airborne pollutants created complex chemical reactions. The problem, however, shows up at the ozone layer.
- In organizations, you can watch the boundary to the system. When a problem occurs, it is likely to show up there.

Patterns of Interaction

When ants switch roles from nest maintenance workers to foragers, they switch without being told to. It is the pattern of interactions, not any particular interaction, that causes them to switch roles. In other words, when the nest maintenance workers interact with enough foragers coming back with bags of groceries, they'll switch.

Minnesota wolves will mark their territory through urination (luckily, "marking territory" is a behavior of lower-order mammals, and this behavior is *never* seen in organizations or societies). Wolf packs create their own territory. The territory is so well-known that if a deer is wounded in the territory of a wolf pack, and it wanders 50 feet into the neighboring wolf territory and dies there, the pack will not follow it. Further, wolves howl to advertise their position to other wolf packs. This ensures that the packs know where each other are, to keep from undesired meetings. (Interestingly, biologists have experimented with this property idea. They have sedated wolves, and dropped them 40 miles from their own territory. The wolf will find its way back to its own territory in a couple days). This pattern of interaction protects the integrity of the wolf packs and their property. No single marking or single howl tells them anything; it is the pattern of interaction that keeps the integrity of the wolf pack boundaries.

The pattern of interactions in an organization will tell you important information about the system. If you see enough interactions, you'll get a sense of the organization's emotional field, which will tell you about the health and fitness of the organization.

You can see patterns of interaction through a variety of methods.

- How do people make decisions? Remember, you're not looking at what people decide, but rather, how they decided it.
- What kinds of things do people talk about when they're at the water cooler (when they're relaxed and unguarded)?
- What motivates policy, procedures, and memos? You're not concerned with, "what *are* the policy and procedures," but rather "what motivates them to put a policy in place at this particular time?"
- What patterns are repeated across the organization? Is peace-mongering repeated over and over in the organization? Poorly differentiated decisions? Chameleon behavior?

- What happens in the organization when someone tries to act in a differentiated manner? How does the system respond?
- Who is the canary in the coal mine? Who is the first to go under pressure? Why?
- What do people say about other people? Remember, you're not looking for any particular comment, you're looking for the patterns.

Attractors, Fractals, and Scaling

Systems are drawn to attractors. An attractor is the preferred position of the system. Adaptive systems tend to move toward the attractors, and then, in the absence of other factors, will stay at the attractor (Lucas, 1997). The journey from the starting point to the attractor is called the *basin of attraction*.

There is cool fractal art at http://projekt.pinknet.cz/fractal/

Organizations have attractors. The attractor draws the attention of the system.

- An attractor may be the values of the leaders of the organization.
- It's possible, albeit rare, that an attractor would be the mission statement of the organization. Mission statements are written for a variety of reasons, and they rarely inspire and stimulate the members of the organization.
- An attractor might be the fear of getting sued. Organizations with this attractor would spend an inordinate amount of time with liability and insurance issues.
- An attractor might be a "peace at all costs" mentality. What matters for peace mongers is "how things look," rather than how things really are.
- An attractor might be a person. If the person were to leave, the organization may become demoralized and could start to disintegrate.
- A budget can easily become an attractor. When this happens, people make decisions based on "if we have the money," rather than making decisions based on "if the idea is a good one or not."

The important clue to determine an organization's attractor is to listen to the direction of people's conversation. It's not what people say is important, but rather, it is the point of their conversations.

A fractal is a type of attractor. A fractal is a computer-generated geometric design that repeats. You see one design when you look at the whole. But when you look at a part of the whole, you see the same design. And when you look at a part of the part, you see the same design as before. And if you took one tiny piece of the design and magnified it, you would still see the same design—the same pattern. This repeating system is also called *scaling* (Olson and Eoyang, 2001).

We see scaling in organizations when there are similar patterns of interaction across the organization. Communication styles, leadership styles, and patterns of work behavior tend to repeat across systems. When we are attentive to scaling, we can better understand organizations and how they function.

Scaling is one of the reasons that leadership is so important in organizations. Differentiated leadership will facilitate differentiated interactions across the system. By the same token, however, poorly-differentiated leadership will help to create a messy system that dysfunctions all over the place.

Chapter 21

Organizations as Complex Adaptive Systems

L eaders don't wake up one day and say, "Today, I'm going to start making my organization a complex adaptive system." The fact is, organizations are complex adaptive systems. You can't decide to make your organization a complex adaptive system; it already is.

What the leader can do is put into place procedures, patterns, and policies that take advantage of that fact. The leader can also put into place structures that hinder the natural tendencies of the system. The leader can use the natural tendencies of systems or the leader can fight the natural tendencies.

- The leader can create environments that facilitate a self-organizing capacity, or the leader can create environments that hinder the self-organizing capacity.
- The leader can work to keep the boundaries permeable when their information and resources must be exchanged with the environment. The leader can keep boundaries impermeable when there is a threat. On the other hand, the leader can be inattentive to boundaries and run the risk of enfeebling the organization.
- The leader can keep the organization attentive to learning, or the leader can ignore learning and hope things work as they always have.
- A leader can put procedures into place that facilitate nimbleness and adaptability, or the leader can put rigid structures into place that keep things the same.

What is at stake is the fitness of the organization. Is the organization a good fit with the environment or is it a poor fit? If it is a poor fit, it will take tremendous energy and massive resources to keep the organization viable. If the organization is a good fit with the environment, a symbiosis—a harmony—will form, keeping the organization viable without a massive influx of resources.

Figure 9
Characteristics of a Poorly-Differentiated Organization

A poorly-differentiated organization will have...	Which means...	And so you'll have to...
High anxiety	Everybody worries about what might happen	Not be drawn into their worries; take chances; be playful; be irreverent
Strong forces of tegetherness	People who want to think alike	Frame situations in new ways
Many triangles	Lots of conflicted relationships; lots of people drawn into other conflicts	Be aware of when triangles are forming and don't allow yourself to be drawn in
A tendency to define other	Blaming, name-calling, mind-reading	Define yourself
Invasiveness	Nosy people who don't mind their own business	Respect other people's boundaries, and draw boundaries for others
Rigidity	Leaders trying to control policies, procedures, and interactions	Gently push boundaries
Either heavy hierarchical control or no leadership	Either rigidity or randomness	As much as possible, discover or create some patterns in your sphere of influence

Further, the economic environment in which your organization finds itself is a complex adaptive system. The market is a complex adaptive system, par excellence. The market will always adapt. It does not show remorse. It does not care if your organization goes bankrupt. The market environment will adapt and change with the tide. If your organization ignores that environment, it will find itself becoming smaller and less relevant on its inevitable path to extinction.

Chapter 22

Principles to Understand Complex Adaptive Systems

1. *Listen, learn, adapt, or die.*
    Environments change, and so species that do not adapt to that environmental change will die. If you're an ant, and you discover a fungus in your anthill, you'll die if you don't pack your bags and move to a new anthill. We adapt by learning. We learn only when we listen. So, organizations that listen to their environment, learn and adapt, will probably thrive and survive in the new environment. Organizations that rigidly stand firm, believing "this is the way we've always done things," will be less fit in the changing environment.

    Unlike most ants, the leaf-cutter ants of the Amazon Basin eat fungus. They actually cultivate and grow their own fungus. Coincidentally, many people's feet do the same thing.

2. *Your organization is organized perfectly to get the results you get.*
    Believe it or not, your organization functions perfectly. It is structured just right to get the results you get. Those results may not be the results you want, but they are the results for how the system has been designed.

    People used to talk about dysfunctional systems. There really are no dysfunctional systems, in the sense that the organization isn't doing what its supposed to. Systems function exactly as we set them up to function. If we want something different, we may have to change the interaction, process, or structure.

3. *You can't change a system, you can only annoy it.*
   This principle is a difficult one. Everybody wants to change the system. Everybody wants to change everybody else. Lots of well-meaning individuals think "everything would be fine if people would just see things the way I see them."

   Here's a news flash: You can't change the system. You can't control it. You can't motivate the system. You can't create something out of nothing. You can't motivate the unmotivated. You can't teach those who don't want to learn. You can't do it. If you do try it, you're just trying to play God. Stop it.

   Let me try to clarify further. You can't change the system *through the force of your own will*. There are ways to change the system, but those are not the ways we usually think. Differentiated behavior brings change to the system. An attention to process, rather than product, can bring change to the system. Changing the patterns of interaction (this is not the same as organizational restructuring) can bring change to the system.

4. *Adaptability and sustainability are the keys to fitness.*
   If an organization wants to do more than just barely survive, a couple capacities are important. The first is the capacity to be adaptive, to develop new characteristics as the environment changes. Second is the capacity to sustain new adaptive changes.

   Not all change is adaptive change. Any organization can change. Adaptive change is a change that makes the organization more fit to the new environment. Typically, in a biological species, a wide number of changes and mutations occur. The changes that make a better fit with the environment survive, the rest die out. The same is true of organizations. A wide variety of changes can happen in an organization. Some organizations are adaptive, some are not. The key is for the leadership to be attentive to which changes are adaptive—which ones will make a better fit with the environment. The other changes should be allowed to die out.

5. *It's not what you don't know that hurts you; it's what you do know that you ignore.*
   Consultants are brought into organizations regularly to help diagnose and plan. The great secret among consultants is that

they get paid to tell the organization what the organization already knows. The people in the organization almost always know what the consultant will tell them. But the people in the organization, for whatever reason, are choosing to ignore the problem or hoping the problem will go away.

6. *If you assume that the problem you're facing is containable, predictable, and linear, you have probably already failed at solving it.*

   In a complex system, problems are complex. You almost never look at one problem—you look bundles of problems that are hopelessly ensnarled together. By definition, an action in one part of a complex adaptive system will have cascading actions throughout the system. One small action will ricochet around the system in nonlinear ways. If you take an action without understanding that a variety of side effects will occur, then you'll be in for some surprises. What's more, if there is a time delay between the action and the side effect, people will forget that the action caused the side effect—further muddling the issue!

7. *The pattern of interactions, not the interactions themselves, are the main thing.*

   Ants know what to do based on the patterns of interaction. They know to switch jobs, recruit themselves for another activity, etc., not because an ant king told them so, but due to the pattern of interaction. It is the pattern of interaction, not the interaction itself, that is the critical thing. Dialog and interaction, not individual traits and characteristics, are the meat and potatoes of how organizations function.

> *An understanding of complex systems will not make decisions easier. Decisions will actually become more difficult, as you take into account multiple variables and nonlinear interactions.*

   If you want to understand an organization, you will want to understand the organization's critical variables: the homeostatic level, the level of anxiety, the ability to be adaptive, and the capacity for differentiated activity. You can't see these things directly. However, when you watch the patterns of interactions, you will receive a variety of clues that will give you information about these critical variables.

8. *There are no right or wrong decisions. There are only tradeoffs.*
Managers spend a lot of time trying to decide if a particular
action is a good decision or a bad decision. Managers, to their
credit, want to make the "right" decision and don't want to make
the "wrong" decision.

   However, if we believe that a system is complex, then we
will realize that an action will always have a variety of side-
effects that cascade throughout the system. There are a variety of
actions and trade-offs in every decision. Every decision we make
and action we take will have effects: some helpful, some not
helpful, some expected some unexpected. There is no perfect
action that will only do the desired effect in some pristine
manner, with no side effects or by-products. Our challenge, when
we are contemplating an action, is to try to assess the potential
side effects of each possible avenue of action, and try to
determine the livability of each side effect.

9. *Diversity increases fitness.*
In biological systems, if the species is spread out over a broad
geographic area, it will tend to develop into subsystems, each
with their own strengths and weaknesses (biologists call this
tendency *genetic drift*). This emergence of diversity protects the
species from catastrophe.

   In organizations, there are many pressures to standardize, but
retaining diversity protects the organization. If organizations put
all their eggs into one basket, the basket could, and eventually
will, fall apart.

   Further, there is often a pressure to "think alike." This is the
forces of togetherness, driven by anxiety, and anxious to think
alike, look alike, and talk alike. A diversity of ideas is the enemy
of the forces of togetherness.

10. *You can fight your organization's CAS tendencies, or you can
take advantage of the CAS tendencies.*
In a traditional organization, the job of the leader is to give
orders, make policies, and to control and supervise. However, in a
complex adaptive system, leadership is dispersed and close to the
front line (as opposed to isolated, away from the front lines). This
position of leadership (dispersed, low, and usually informal)
enhances the survivability of the system.

In a complex adaptive system, the leader is still the leader—but the tasks are different. In a complex adaptive system, the job of the leader is to keep the community attentive to learning. Rather than trying to control people, the environment, or the future (control is a myth, anyway), the leadership of a complex adaptive system defines the boundary, keeps the community learning, and empowers the people to make the decisions they need to enhance the sustainability and adaptability of the system.

Humans, with a few exceptions, are not ants. Contrary to ant life, human organizations will always have a leader.

Forced behavior, as we see in hierarchical organizations, reduces the capacity of the system to be adaptive. So, if a leader is making decisions all the time for people, they shouldn't be surprised that they have to make decisions all the time for people!

Chapter 23

Summary

The study of complex adaptive systems is not simply a metaphor for organizational behavior. There are tendencies and standard ways that all systems function—whether they are biological or social systems.

Once we realize the implications of the idea that an organization is a complex adaptive system, then we can make some intentional choices. We can choose to fight all the natural tendencies of the system. Many managers do that (and never know that is what they are doing).

Or, we can take advantage of the natural dynamics of complex adaptive systems. We can keep the organization attentive to learning; we can influence boundaries; we can understand the organization better by looking at the patterns of interaction; we can encourage self-organization; and we can make adapting a way of life.

Free Vampire Hunting Tips

Handling poisonous snakes is bad enough. But handling vampires is worse.

The problem of course, is that a snake will only bite you if you step on it. A vampire goes out searching for blood. Vampires, like snakes, operate only out of instinct. Vampires are driven—not by reason or common sense—but only by the burning thirst inside them. Vampires don't bargain. Trying to educate them is useless. Giving them reasons why they shouldn't bite you is futile—it will frustrate you and annoy the vampire.

The only thing that will stop a vampire is a boundary. Garlic on the windows or a cross on the door will stop vampires in their tracks.

In organizations, there are people who are severely instinctual and very poorly differentiated. This is the organizational vampire. The organizational vampire is much worse than the organizational snake. The snake is only dangerous when you step on it. The organizational vampire is dangerous all the time. They are constantly crabby, reactive, defensive, invasive, and predatory. They suck the life-blood (vitality, focus, motivation, fun) out of an organization. Vampires are sometimes suave and sometimes ugly; they are sometimes charming and sometimes revolting; they are sometimes well-mannered and sometimes abrasive. But they never, ever lose their taste for blood.

Some organizations try to reason with the vampire. Most just work around the vampire. They ignore the vampire, or meet most of their demands so they don't become spiteful (this is why many vampires tend to control the agenda of organizations). But you can't change them. They will eventually be back—and as thirsty as the last time. Mind you, it's not personal—it's simply their instinct. Vampires don't care who they chew on—they simply want to eat something. They won't stop with more information, or a better presentation. And you can't change their personality.

The only thing that will stop them is a boundary. Interpersonal and organizational boundaries are garlic to a vampire. The only way to stop them is to say, "you can't cross this line."

This process of boundary-setting is the only thing that will stop an organizational vampire.

By the way, this is a metaphor. There is no such thing as a real vampire. As far as you know.

Cool Words

Adaptive: When components of a system change over time as a result of new influences from the environment. This adapting is usually done without the direction of a central authority.

Amplifying Feedback: Feedback that increases the likelihood of the action reoccurring.

Analysis: Breaking a phenomenon down to its component parts to try to understand it.

Anxiety: Long-term, chronic assessments of dread that frequently motivate personal and organizational behavior. Usually learned in our family of origin.

Attractors: The situation, attitude, or process that draws and centers the organization.

Balancing Process: Processes that hinder change and promote homeostasis.

Boundary: The edge of the system. The line that separates one system from another.

Butterfly Effect: The idea that a small change, like the flap of butterfly wings, can create enormous change as it cascades via nonlinear relationships through the systems.

Chaos: The state that, on first glance, appears to be random and unpredictable. However, with a closer look, there is a pattern.

Closed System: A system that has rigid, impermeable boundaries and shares no information with its environment.

Compensating Feedback: Feedback that hinders change and promotes homeostasis.

Complex Adaptive Systems: Interconnected groups of systems that adapt to become better fit with the environment. This adaptation usually occurs without the direction of a central authority figure.

Complex Evolving System: A system that evolves, rather than adapts.

Complexity: The nature of systemic phenomenon—interconnected and nonlinear.

Complexity Science: The study of complex phenomena. Sometimes this term is used to describe all of the complex systems theories (cybernetics, general systems theory, process adaptive systems).

Damping Feedback: Feedback that promotes homeostasis and hinders change. Also called balancing processes and compensating feedback.

Differentiate: The process of differentiation; to articulate our own goals in the midst of countering opinions.

Differentiation: Perception of boundaries; ability to articulate own goals in the midst of countering opinions. To be aware of the emotional field without being controlled by it.

Direct Adaptation: Forced change by an authority figure.

Diversity: Many different types of relationships within the system.

Emotional Barrier: Inability to think about a topic. A stunted imagination.

Emergence: The quality of complex phenomena to have a new character emerge that is different than the individual parts.

Emotional Triangles: A conflicted relationship that has drafted a third person (or place or thing) to dissipate the anxiety of the conflict.

Feedback: Information that comes back to the system.

Fitness: The quality of being sustainable. Being able to thrive in a particular environment.

Forces of Togetherness: The push in an organization to think alike and act alike. The more anxiety there is in the system, the more likely the members will try to dilute the anxiety by desiring sameness.

Fractal: A cool computer-generated design that repeats a pattern in large and small ways.

Homeostasis: The systemic process of keeping things the same.

Identified Patient: The scapegoat for a conflicted system—usually where the symptoms erupt.

Impermeable Boundary: A boundary that will not let anything in or out.

Indefinite Number of Agents: The characteristic of complex systems that suggests that there are so many influences on the system that they would be impossible to count.

Indistinct Boundary: A boundary so permeable it will allow any information or stimulus through.

Interdependence: The reliance on all the parts of the system to achieve maximum effectiveness.

Learning: A change in behavior for a particular reason.

Linear: Cause and effect. A process that is completely obvious. Simplistic and predictable. See "politician."

Linear Thinking: Simplistic, cause-effect thinking.

Mechanization: The view of the world suggesting everything runs like a machine; every effect has one cause.

Multiple Causation: The view that every effect has several linear causes; this is a more sophisticated type of linear thinking.

Nonlinear Relationships: Relationships and processes that are not cause and effect. Particular actions cascade and ricochet around the system in an unpredictable manner.

Natural Selection: The biological principle that suggests that systems that are not a good fit to their environment will die off.

Open System: A system with permeable boundaries, exchanging information with agents in the environment.

Organizational Learning: The process that an organization uses to learn from its experience and its environment, with the intention of adapting and making changes for enhanced improvement and effectiveness.

Peace-Mongering: A peace-at-all-costs mentality. An intentional, willful suppression of conflict for the purpose of giving the illusion of peace.

Permeable Boundary: A boundary on a system that allows some information, stimulus, and material through itself, and keeps some stuff out.

Play: An attitude that values creativity, innovation, and spontenaity. It is an antidote for organizational stuckness.

Process Adaptive Systems: A particular model or school of thought in complexity science. This book outlines process adaptive systems.

Reductionism: The view that the whole can be understood by analyzing the parts.

Scaling: The tendency for a system to repeat behaviors on many levels and in many places. Not only behaviors are repeated, but patterns of interaction and systemic dynamics are repeated.

Secrets: Intentionally keeping information from people for the purpose of gaining influence over them.

Self-Organize: Readjusting and adapting without the direction of a central authority figure.

Sustainability: The quality of being fit with the environment over the long term.

Symptom: The visible, outward manifestations of a problem.

System: A set of interdependent, interactive parts, working together for some purpose.

Systems Thinking: A way of observing and perceiving that uses systems as the conceptual framework for understanding how things work.

Transdisciplinary: Taking insights and principles from a variety of academic disciplines, rather than being housed in one discipline.

Wormic: Of or relating to worms.

Bibliography, References, and Further Study

Books

Ackoff, R. (1991). *Ackoff's fables: Irreverent reflections on business and bureaucracy*. New York: John Wiley & Sons, Inc.

Anderson, V. & Johnson, L. (1997). *Systems thinking basics: From concepts to causal loops*. Cambridge, MA: Pegasus Communications, Inc.

Axelrod, R. & Cohen, M. (1999). *Harnessing complexity: Organizational implications of a scientific frontier*. New York: Free Press.

Baskin, K. (1998). *Corporate DNA: Learning from life*. Boston, MA: Butterworth Heinemann.

Bertalanffy, L. (1969). *General systems theory: Foundations, development, applications*. New York: George Braziller, Inc.

Bowen, M. (1978). *Family therapy in clinical practice*. Northvale, NJ: Jason Aronson Publishers.

Briggs, J. & Peat, F.D. (1999). *Seven life lessons of chaos: Timeless wisdom from the science of chance*. New York: Harper Collins.

Capra, F. (1996). *The web of life: A new scientific understanding of living systems*. New York: Doubleday Dell Publishing Company.

Chawla, S. & Renesch, J. (Eds.). (1995). *Learning organizations: Developing cultures for tomorrow's workplace*. Portland, OR: Productivity Press.

Clippinger, J.H. (Ed.). (1999). *The biology of business: Decoding the natural laws of enterprise*. San Francisco: Jossey-Bass.

Comella, P., Bader, J., Ball, J., Wiseman, K., & Sagar, R. (1996). *The emotional side of organizations: Applications of Bowen theory*. Washington D.C.: Georgetown University Family Center.

Coveney, P. & Highfield, R. (1995). *Frontiers of complexity: The search for order in a chaotic world.* New York: Fawcett Columbine.

Eoyang, G. (1997). *Coping with chaos: Seven simple tools.* Cheyenne, WY: Lagumo Corp.

Friedman, E. (1985). *Generation to generation: Family process in church and synagogue.* New York: Guilford Press.

Friedman, E. (1990). *Friedman's fables.* New York: Guilford Press.

Friedman, E. (1996). *Reinventing leadership: Change in an age of anxiety.* [video and discussion guide]. (Available from Guilford Press, 72 Spring Street, New York, NY 10012).

Friedman, E. (1999). *A failure of nerve: Leadership in the age of the quick fix.* Bethesda, MD: The Edwin Friedman Estate.

Fulmer, W. (2000). *Shaping the adaptive organization: Landscapes, learning, and leadership in volatile times.* New York: American Management Association.

Gell-Mann, M. (1994). *The quark and the jaguar: Adventures in the simple and the complex.* New York: W.H. Freeman and Company.

Gleick, J. (1987). *Chaos: Making of a new science.* New York: Penguin Books.

Gordon, D. (1999). *Ants at work: How an insect society is organized.* New York: The Free Press.

Haeckel, S. (1999). *Adaptive enterprise: Creating and leading sense-and-respond organizations.* Boston, MA: Harvard Business School Press.

Haines, S.G. (1998). *The manager's pocket guide to systems thinking and learning.* Amherst, MA: Centre for Strategic Management.

Hanson, B.G. (1995). *General systems theory beginning with wholes.* Washington, DC: Taylor and Francis Publishers.

Holland, J. (1995). *Hidden order: How adaptation builds complexity.* Cambridge, MA: Perseus Books.

Kauffman, D. (1980). *Systems one: An introduction to systems thinking.* Minneapolis, MN: Future Systems, Inc.

Kauffman, S. (1995). *At home in the universe: The search for the laws of self-organization and complexity.* New York: Oxford University Press.

Kelly, S. & Allison, M.A. (1999). *The complexity advantage: How the science of complexity can help your business achieve peak performance.* New York: McGraw Hill.

Kerr, M. & Bowen, M. (1988). *Family Evaluation.* New York: W.W. & Norton Co.

Kim, D. (1994). *System archetypes one: Diagnosing systemic issues and designing high-leverage interventions.* Cambridge, MA: Pegasus Communications, Inc.

Kim, D. (1995). *Systems thinking tools: A user's reference guide.* Waltham, MA: Pegasus Communications.

Kim, D. (1999). *Introduction to systems thinking.* Cambridge, MA: Pegasus Communications, Inc.

Kim, D. & Anderson, V. (1998). *Systems archetype basics: From story to structure.* Waltham, MA: Pegasus Communications, Inc.

Lewin, R. & Regine, B. (2000). *The soul at work, listen, respond, let go: Embracing complexity science for business success.* New York: Simon and Schuster.

Lewin, R. (1999). *Complexity: Life at the edge of chaos.* Chicago, IL: University of Chicago Press.

Marshall, I. & Zohar, D. (1997). *Who's afraid of Schrodinger's cat?* New York: William Morrow and Company.

Mayr, E. (1997). *This is biology: The science of the living world.* Cambridge, MA: Belknap Press of Harvard University Press.

Moore, J. (1997). *The death of competition: Leadership and strategy in the age of business ecosystems.* New York: HarperCollins.

Moring, G. (2000). *The complete idiot's guide to understanding Einstein.* Indianapolis, IN: Macmillan USA, Inc.

O'Conner, J. & McDermott, I. (1997). *The art of systems thinking: Essential skills for creativity and problem solving.* San Francisco, CA: Thorsons, an Imprint of HarperCollins Publishers.

Olson, E. & Eoyang, G. (2001). *Facilitating organizationa change: Lessons from complexity science.* San Francisco, CA: Jossey-Bass Publishers.

O'Reilly, K.W. & Johnson, L. (Eds.). (1998). *The new workplace: Transforming the character and culture of our organizations.* Waltham, MA: Pegasus Communications.

Papero, D. (1990). *Bowen family systems theory.* Needham Heights, MA: Allyn and Bacon.

Ramsey, P. (1997). *Billibonk and the thorn patch*. Waltham, MA: Pegasus Communications.

Richmond, B. (2000). *The "thinking" in systems thinking: Seven essential skills*. Waltham, MA: Pegasus Communications.

Sagar, R.R. (1997). *Bowen theory and practice: Feature articles from the family center report, 1979-1996*. Washington D.C.: Georgetown University Family Center.

Sagar, R.R. & Wiseman, K.K. (Eds.)(1982). *Understanding organizations: Applications of Bowen family systems theory*. Washington D.C.: Georgetown University Family Center.

Sanders, T. I. (1998*). Strategic thinking and the new science: Planning in the midst of chaos, complexity, and change*. New York: The Free Press.

Satir, V. (1972). *Peoplemaking*. Palo Alto, CA: Science and Behahavior Books.

Senge, P. (1990). *The fifth discipline: The art and practice of the learning organization*. New York: Doubleday.

Senge, P. (1994). *Building learning infrastructure* (Cassette Recording No. T9404). Waltham, MA: Pegasus Communications.

Senge, P., Kleiner, A., Roberts, C., Ross, R., & Smith, B. (1994). *The fifth discipline fieldbook: Strategies and tools for building a learning organization*. New York: Currency Doubleday.

Senge, P., Kleiner, A., Roberts, C., Roth., G., & Ross, R. (1998) *The dance of change: The challenges of sustaining momentum in learning organizations*. New York: Doubleday.

Senge, P., Cambron-McCabe, N., Lucas, T., Smith, B., Dutton, J., & Kleiner, A. (2000). *Schools that learn: A fifth discipline fieldbook for educators, parents, and everyone who cares about education*. New York: Doubleday.

Stacey, R. (2001). *Complex responsive processes in organizations: Learning and knowledge creation*. New York: Routledge.

Waring, A. (1996). *Practical systems thinking*. Boston, MA: International Thomson Business Press.

Wheatley, M. (1992). *Leadership and the new science: Learning about organizations from an orderly universe*. New York: Berret-Koehler.

Zabarenko, D. (2000, July 20). *Getting there faster: Light's speed accelerated* [Electronic version]. Reuter's.

Periodicals

Advances in Complex Systems
http://www.santafe.edu/~bonabeau/

Complexity Digest
http://www.comdig.org/
Complexity International
http://www.csu.edu.au/ci/

Emergence
http://www.emergence.org/

Family Systems
http://www.georgetownfamilycenter.org/

Systems Thinker Newsletter
http://www.pegasuscom.com/tstpage.html

Systemic Practice and Action Research
http://www.wkap.nl/journalhome.htm/1094-429X

Organizations

American Society for Cybernetics
http://www.asc-cybernetics.org/

Bowen Center for the Study of the Family
http://www.georgetownfamilycenter.org/

Chaordic Alliance
http://www.chaordic.org/

Georgetown Family Center
http://www.georgetownfamilycenter.org/

International Society for the Systems Sciences
http://www.isss.org/

New England Complex Systems Institute
http://necsi.org/

Pegasus Communications
http://www.pegasuscom.com/

Principia Cybernetica
http://pespmc1.vub.ac.be/DEFAULT.html

Santa Fe Institute
http://www.santafe.edu/

Society for Chaos Theory and Life Sciences
http://www.socictyforchaostheory.org/

Society for Organizational Learning
http://www.solonline.org/solonline/

System Dynamics Society
http://www.albany.edu/cpr/sds/

Index

About the Authors

Jim Ollhoff is the Associate Dean of Human Services at Concordia University, St. Paul, Minnesota. Jim's academic background includes education, family studies, social psychology, and management. Jim is hoping to finish his PhD dissertation before dying of old age. He lives in Farmington, Minnesota, in a system with a wife, a son, two cats, and a dog. He can be reached at: ollhoff@charter.net

Michael Walcheski is the Chair of the Department of Family Studies at Concordia University, St. Paul, Minnesota. He has a PhD in Marriage and Family Therapy and is a Licensed Marriage and Family Therapist. Michael lives in Roseville, Minnesota in a system with a wife, two children and a snotty cat. He can be reached at: walcheski@unique-software.com